SAGE was founded in 1965 by Sara Miller McCune to support the dissemination of usable knowledge by publishing innovative and high-quality research and teaching content. Today, we publish over 900 journals, including those of more than 400 learned societies, more than 800 new books per year, and a growing range of library products including archives, data, case studies, reports, and video. SAGE remains majority-owned by our founder, and after Sara's lifetime will become owned by a charitable trust that secures our continued independence.

Los Angeles | London | New Delhi | Singapore | Washington DC | Melbourne

ENVIRONMENTAL ACCOUNTING, SUSTAINABILITY & ACCOUNTABILITY

ENVIRONMENTAL ACCOUNTING, SUSTAINABILITY & ACCOUNTABILITY

SOMNATH DEBNATH

Los Angeles | London | New Delhi
Singapore | Washington DC | Melbourne

Copyright © Somnath Debnath, 2019

All rights reserved. No part of this book may be reproduced or utilized in any form or by any means, electronic or mechanical, including photocopying, recording, or by any information storage or retrieval system, without permission in writing from the publisher.

First published in 2019 by

SAGE Publications India Pvt Ltd
B1/I-1 Mohan Cooperative Industrial Area
Mathura Road, New Delhi 110 044, India
www.sagepub.in

SAGE Publications Inc
2455 Teller Road
Thousand Oaks, California 91320, USA

SAGE Publications Ltd
1 Oliver's Yard, 55 City Road
London EC1Y 1SP, United Kingdom

SAGE Publications Asia-Pacific Pte Ltd
18 Cross Street #10-10/11/12
China Square Central
Singapore 048423

Published by Vivek Mehra for SAGE Publications India Pvt Ltd. Typeset in 10.5/13 pt Berkeley by Zaza Eunice, Hosur, Tamil Nadu, India.

Library of Congress Cataloging-in-Publication Data Available

ISBN: 978-93-532-8464-0 (HB)

SAGE Team: Abhijit Baroi, Manisha Mathews, Ankit Verma and Rajinder Kaur

If many small people in many small places change in a small way, the face of the earth changes.

—An African proverb

Thank you for choosing a SAGE product!
If you have any comment, observation or feedback,
I would like to personally hear from you.

Please write to me at **contactceo@sagepub.in**

Vivek Mehra, Managing Director and CEO, SAGE India.

Bulk Sales

SAGE India offers special discounts
for purchase of books in bulk.
We also make available special imprints
and excerpts from our books on demand.

For orders and enquiries, write to us at

Marketing Department
SAGE Publications India Pvt Ltd
B1/I-1, Mohan Cooperative Industrial Area
Mathura Road, Post Bag 7
New Delhi 110044, India

E-mail us at **marketing@sagepub.in**

Subscribe to our mailing list
Write to **marketing@sagepub.in**

This book is also available as an e-book.

CONTENTS

List of Figures	ix
List of Tables	xi
List of Abbreviations	xiii
Preface	xix
Acknowledgements	xxiii

Chapter 1	Sustainability and Accounting Sciences: Two Independent Paradigms	1

Section I: Accounting and Accountability: Traditional Paradigm

Chapter 2	Organizational Theories and Accountability	25
Chapter 3	Financial Accounting, Reporting and Accountability	49
Chapter 4	Cost and Managerial Accounting: Supporting Management beyond Numbers	72
Chapter 5	Economics: A Rational Argument to Ignore Environment	86

Section II: Environment and Accounting Theories: Contemporary Advances

Chapter 6	Environmental Considerations and Conventional Accounting Theories	101

Chapter 7	Contemporary Developments in Green(ing) Accounting	119
Chapter 8	Methodological Developments in Environmental Management Accounting	139
Chapter 9	Advances in Other Environmental Frameworks	159

Section III: Environmental Accounting: A Dimensional View of Accounting

Chapter 10	Environmental Accounting: Connecting Critical and Normative Theories of Accounting	173
Chapter 11	Environmental Accounting: An Independent Accounting Viewpoint	185
Chapter 12	Advancements in Costing Models to Handle Externalities	208
Chapter 13	Environmental Accounting: Part I	226
Chapter 14	Environmental Accounting: Part II	243

Section IV: Accounting Sciences and Sustainability Theories: Managerial Implications and Recent Advances

Chapter 15	Environmental Accounting and Managerial Implications I: Carbon Accounting	271
Chapter 16	Environmental Accounting and Managerial Implications II: Other Advances	290
Chapter 17	Environmental Management Systems and Greening Firms	307
Chapter 18	Sustainability and Environmental Interfaces: Recent Advances	330

Appendix: Mathematical Modelling of Complex Waste	346
References	362
Index	388
About the Author	396

LIST OF FIGURES

6.1	Accounts information life cycle	106
10.1	Accounting paradigms of sustainability as part of accounting frameworks	183
11.1	Accounting implications of a single transaction in parallel accounting universes	193
11.2	Environmental accounting framework	198
12.1	Energy consumption and GHG emissions generated as part of MSWM	213
12.2	Flow of aspects and impact tree	219
12.3	Waste flow through material (re)cycle	222
12.4	Layered nature of environmental impacts and costs	224
13.1	Operational layout of FPP	228
13.2	Materials and waste flow	233
13.3	Energy flow	234
13.4	Water flow	235
14.1	Operational layout of CHS	246
17.1	Functional decomposition of a business function	317
17.2	Integrated waste management system	327

LIST OF TABLES

8.1	Comparative characterization of different regional indices	151
8.2	Environmental performance indicators (Examples)	155
11.1	Comparative accounting viewpoints for sample business transactions	195
11.2	Comparison of environmental accounting with other frameworks	197
11.3	Transactional interrelationship within accounting dimensions	200
12.1	MSWM activities and environmental contributions	214
13.1	Summarized flow of costs on account of waste in FPP	230
13.2	EMA computations for FPP	236
13.3	Select environmental aspects of FPP	238
13.4a	Solid waste (externality) t-account	240
13.4b	Product B: Total environmental aspects t-account	241
13.5	Externalities incurred/saved due to business activities of FPP	242
14.1	Periodic consumption of resources in respective units	248
14.2	Environmental aspects from resource consumption (with units)	249
14.3	EMA computations for CHS	250
14.4	Analysis of waste generated	251
14.5a	Solid waste (externality) t-account	252
14.5b	Waste water t-account	253
14.5c	Emissions t-account	254
14.5d	Environmental liability t-account	255

14.6	Externalities incurred/saved due to the business activities of CHS	256
14.7	Travel miles and emissions produced during Project A	259
14.8	T-account of emission aspects due to travel miles in Project A	261
15.1	Step-down allocation of aggregate GHG to different processes (in absence of GHG profiles)	280
15.2	GHG profiling of organizational processes (using the proposed framework)	282
15.3	Domestic dispatch (in tonnes)	283
15.4	Export dispatch (in tonnes)	284
15.5	Carbon accounting, leveraging environmental accounting	287
17.1	Different EMS certification standards	313
17.2	Eco-friendly organizational measures and their positive impacts	322
A1	Dispersion of FA aspects in decreasing order of specificity	355

LIST OF ABBREVIATIONS

AAAJ	*Accounting, Auditing & Accountability Journal*
AAUs	allowable accounting units
ABC	activity-based costing
ABM	activity-based management
AHP	analytical hierarchy process
AICPA	American Institute of Certified Public Accountants
AIS	accounting information system
ASC	Accounting Standards Codification (prevalent in the US)
ASQ	American Society for Quality
BREEAM	Building Research Establishment Environment Assessment Method
BRIC	Brazil, Russia, India, China
CAGR	compound annual growth ratio
capex	capital expenditure
CASBEE	Comprehensive Assessment System for Built Environment Efficiency
CBA	cost–benefit analysis
CDI	city development index
CDM	Clean Development Mechanism (as defined by the Kyoto Protocol)
CDP	Carbon Disclosure Project
CER	certified emission reduction (units)
CERES	Coalition for Environmentally Responsible Economies
CIP	construction-in-progress
CMA	carbon management accounting
COP	Conference of the Parties (to the UNFCCC)
CSI	city sustainability index

CSR	corporate social responsibility
CVP	cost–volume–profit (analysis)
DOI	diffusion of innovation
EASO	environmental assessment of sites and organizations
EBITDA	earnings before interest, taxes, depreciation and amortization
ECA	environmental cost accounting
ECEA	environmental capability enhancing asset
EEA	European Environment Agency
EF	ecological footprint
EIA	environmental impact assessment
EICI	eco-intensity change index
E-LCC	environmentally sensitive life-cycle costing
EMA	environmental management accounting
EMAS	Eco-Management and Audit Scheme (EU)
EMS	environmental management system
ENRAP	Environmental and Natural Resources Accounting Project (Philippines)
EPA	Environmental Protection Agency
EPC	engineering, procurement and construction
EPIs	environmental performance indicators
ERP	enterprise resource planning
ESI	environmental sensitivity index
ETP	effluent treatment plant
ETS	Emissions Trading Scheme (EU)
EU	European Union
EUA	European Union Allowance (EU)
FASB	Financial Accounting Standards Board (of the USA)
FCA	full-cost accounting
FDF	Finnish Defence Forces
FIFO	first in, first out
GAAPs	generally accepted accounting principles
GBTool	Green Building Tool
GCV	gross calorific value
GDP	gross domestic product
GE	General Electric (company)
GHGs	greenhouse gases
GRI	Global Reporting Initiative
GSCM	green supply chain management
HACCP	hazard analysis and critical control points

HDI	human development index
HERS	Home Energy Rating System
HFC	hydrofluorocarbon
HLRW	high-level radioactive waste
HR	human resources
HSD	high speed diesel
IAS	International Accounting Standards
IASB	International Accounting Standards Board
IASC	International Accounting Standards Committee
ICAC	Instituto de Contabilidad y Auditoría de Cuentas
IFAC	International Federation of Accountants
IFRS	International Financial Reporting Standards
IIRC	International Integrated Reporting Council
IOA	input–output analysis
IPCC	Intergovernmental Panel on Climate Change
ISO	International Organization for Standardization
IMF	International Monetary Fund
IMU	Institut für Management und Umwelt (Germany)
IUSIL	international urban sustainability indicators list
IWM	integrated waste management
JI	joint implementation (as defined by the Kyoto Protocol)
LA21	Local Agenda 21
LCA	life-cycle analysis
LCC	life-cycle costing
LEED	Leadership in Energy and Environmental Design
LFG	landfill gas
LIFO	last in, first out
LLRW	low-level radioactive waste
LPI	living plant index
LULUCF	land use and land use changes and forestry
MCA	multi-criteria analysis
MCDA	multiple-criteria decision analysis
MCDM	multi-criteria decision-making (technique)
MDA	multiple discriminant analysis
MES	manufacturing execution function
METI	Ministry of Economy, Trade and Industry (Japan)
MFA	material flow accounting (economic)
MFCA	material flow cost accounting
MIS	management information system

MOE	Ministry of Environment (Japan)
MSW	municipal solid waste
NAMEA	National Accounting Matrix including Environmental Accounts
NPOs	non-product outputs
NPV	net present value
OI	organizational innovation
opex	operational expenditure
PDCA	plan–do–check–act (cycle)
PEST	political, economic, social and technological (framework of analysis)
PESTEL	political, economic, social, technological, environmental and legal (framework of analysis)
PFC	perfluorocarbon
PIOT	physical input–output table
PM	particulate matter (in pollution)
PMO	project management office
ppm	parts per million
PT	process theory (of innovation)
PV	present value
QR	quality control
R&D	research and development
REA	resources, events, agents (model)
ROI	return on investment
SAM	sustainability assessment model
SASB	Sustainability Accounting Standards Board
SDI	sustainability development index
SEA	social and environmental accounting
SEAR	social and environmental reporting
SEC	Securities and Exchange Commission (of the USA)
SEEA	System of Ecological and Environmental Accounting
SMA	strategic management accounting
SMEs	small and medium enterprises
SNA	System of National Accounts (as defined by the UN)
SSN	space surveillance network
SWOT	strengths, weaknesses, opportunities and threats (as framework for analysis)
TBL	triple bottom line
TCA	total cost approach
TCO	total cost of ownership

TOC	theory of constraints
TPD	tonnes per day
UNCSD	United Nations Commission on Sustainable Development
UNDSD	United Nations Division for Sustainable Development
UNEP	United Nations Environment Programme
UNFCCC	United Nations Framework Convention on Climate Change
UOM	unit of measurement
USA	United States of America
USEPA	United States Environmental Protection Agency
WBCSD	World Business Council for Sustainable Development
WI	well-being index
WRI	World Resources Institute
WTE	waste-to-energy
WTP	willingness to pay

PREFACE

Almost a decade ago, I embarked on the journey of understanding how accounting sciences can contribute to environmental concerns and the planetary well-being. Quite frankly, I found these concerns to be challenging to the existing paradigms of accounting sciences and the secure world of isolated views that the financial data and information offer. More so, because the acuteness of accounting framework is inherent to its design, and perseverance to uphold the acuteness is critical to an accounting system. Critical theorists have postulated this as inherent limitation of accounting, which I firmly believe is not challenging to the accounting language per se, but to the viewpoint within which an accounting framework is institutionalized. This, in all fairness, was the main reason why I delved deeper, to establish how the chosen viewpoint restricts the framework from engaging in a conversation meaningful beyond its central theme and eschews innovativeness of outlook, preventing the framework from examining business conduct in unconventional ways.

Another perspective is that the fickle nature of our collective wisdom has lauded some of our long-standing institutions—industries and businesses—for contributing to the progress of civilization and improving our lives, often decorating them and showering accolades upon them, only to denounce them a moment later for not being responsive enough to environmental and societal challenges. If the new societal order expects business enterprises to be addressing some of these concerns, we need to clearly convey our expectations for them to evolve a mechanism to demonstrate their engagement. This contrasts with the focal expectation of markets that only reward superior economic performance. Accordingly, rendition of sustainability as a practicum that grounds our hopes in industries, I honestly believe, mismatches what we collectively intend to achieve even before deliberating how we want to achieve that.

Within these and other complexities, this book explores collective viewpoints that can serve as a reference point through which business and accounting advances towards sustainability can be looked at, and questions whether environmental view can be translated to a unique accounting theme to reflect stakeholders' expectations and capture firm-environment exchange by relying on certain common rules of engagement that can pave the way for its practical adoption.

Based on the expansion of knowledge in the ecological sciences and our ability to identify the anthropocentric impacts on ecology, isolated boundaries of accounting considerations, labelled as dimensionality that frames the accounting sciences, can contribute to the evolution of ecological accounting, freeing accounting language from being subservience to the economic theories and from its morphic resonance to financial accounting, to let its application grow beyond the boundaries of business and economics. More than the pragmatic orientation of such a solution, I believe, any human creation can surpass the bounds of time and space so long as it can convey *some* eternal values or be a part of them. This is where the need for accounting to be grounded in accountability arises—beyond the confines of an accounting process answerable to a few (owners and third parties) and help transcend individual businesses to enable the corporate character they need to be accountable to. Wishful thinking, should I say?

This book is based on contemporary advances in research that are yet to achieve a firm theoretical grounding while leveraging concepts from different disciplines that I sincerely believe would be the future ingredients of our collective thinking as well. I earnestly hope that the concepts explored and evolved here will engage readers and spark off the requisite discussions within businesses, academia and society.

At the same time, I accept responsibility for any mistakes, lapses and omissions, such as disregarding any prominent areas of research and scholarship that may be relevant to the topic but have not been covered due to lack of relevance to the specific line of arguments presented in this book. I look forward to hearing from you—your feedback, thoughts, criticisms and suggestions.

ACKNOWLEDGEMENTS

I wish to acknowledge the contributions of the learned minds and institutions that have furthered the human quest for knowledge, not to mention all individuals who helped me appreciate the interconnectedness of everything we know as life!

Sustainability and Accounting Sciences

Two Independent Paradigms

Trade and commerce have been integral to the evolution of human society and offer a remarkable tale of exploration and innovation, contributing to the complex narrative of societal development, as they themselves evolved into today's human institutions. Trade and commerce have not only connected societies and markets, they have connected different geographies through the flow of goods and services in the past and hopefully will continue to do so in the future as well. Probably that is why they have existed since the dawn of human civilization, connecting people across races and regions, bringing humanity together for better (migration and exploration) and for worse (slave trade and colonization). Trade linguistically relates to exploration like as a course, track or to tread. Commerce, on the other hand, means to come together to trade (in Latin). How old could the activity of trading be? Watson (2009) estimates trade to have existed since prehistoric times, about 150,000 years ago, when barter was the main way of exchanging goods and services, long before currency or other form of exchange were invented. Other than to achieve self-sufficiency, we can easily imagine that trade and commerce pushed brave navigators and travellers to discover new geographies and routes, to invent new modes of transportation and navigational technologies, and to contribute to cultural enrichment by bringing together diverse societies, while satisfying the undying human thirst to venture out, to seek the unknown, and to go where no one has gone before. We acknowledge the efforts of the brave souls who acted as conduits, messengers and representatives of their realms, and as predators and looters at times, who left the comforts of their homelands to brave the dangers of the unknown, lifting the human spirit to its potential.

Bringing the scientific perspective to trade involves looking for evidence that it existed in the prehistoric era—for example, trade between geo-locals may be indicated by tools from a certain era being found in a region without corresponding evidence of their source materials from the same age. Archaeological evidence from the prehistoric era suggests obsidian was the main raw material used for making tools in the Palaeolithic period. There have been several archaeometric studies involving different archaeological sites where obsidian tools from different prehistoric periods have been found, their age having been validated by differences in their material composition, yet without a corresponding source of obsidian of the same age being found within the same region, corroborating trade across regions (Abedil et al. 2018; and other works of Renfrew and Wright (1969) from 1969 to 1980). Here, regions are not to be understood in terms of geo-political boundaries, but more as a conceptual map within which different societal constructs prevailed in a given period. Of course, historical data from recorded history—for example, from the Iron Age onwards—offer much greater detail of evidence for trade in later periods, including the discoveries of land- and sea-trade routes in different periods.

In contrast to trade, commerce constitutes the exchange of goods and services on a much larger scale, mostly at an aggregate level within a region or state. Other than the more organized societal progress that is required for commerce to evolve, the rise of commerce also depended on the integration of systemic elements of society that contributed to its efficiency, operations and scaling, with a need for the relevant infrastructure to be in place.

Another institution that supported trade and commerce and aided the human desire to come together to connect with their own kind in distant places was the market. The markets or marketplaces in different regions supported trade and commerce by acting as a physical hub or space where buyers and sellers could meet and exchange goods and services, not to mention the exchange of ideas and camaraderie, which eventually led to markets evolving into lasting human institutions. In the digital era, we see them evolve into a virtual marketplace, although

I am not sure if this can replace the physical connection that markets traditionally offered.

With the concept of regional boundary firming up through time and the rise of regional sovereignty, commercial practices moved from being socially characterized to become politically determined. This relates to the legitimacy offered through institutional arrangements, including honouring trade and commercial practices through legal frameworks and upholding the legitimacy of commercial dealings in different eras. While differences inherent to the time periods could be an area of scholarly interest, institutional legitimacy contributed to the efficiency of trade and commerce and aided in alleviating difficult conditions due to regional disturbances such as war and famine, as well as in developing redressal mechanisms to address the grievances of trading partners, which otherwise would have been impossible to handle equitably. With the gradual shift in societies to become increasingly connected, we can extrapolate that trade and commerce enriched collective understanding of shared dependencies and enhanced their abilities to deal with surpluses and shortages of commodities.

Although social and anthropological studies support existence of trade and commerce for at least 100,000 years, the efficiency in trade and commerce improved significantly with invention of money, which acted as transaction intermediary and a unit of exchange. Research is yet to establish the details of intermediate forms of exchange that historically existed before the era of commodity money, but if fables can be considered as narratives from a bygone era, the existence of some form of settlements cannot be denied. We can safely assume that there were different types of items that could have potentially played the role of commodity money, such as weights or counts of a common produce from a region or unique items commonly available throughout a region. Examples of this type of money could be bushels of barley, a definite volume of rice, counts of shells,[1] and so on.

---

[1] The use of seashells of a specific kind to pay for the last journey of dead relatives to their heavenly abode is a fable that has survived through ages, where the boatman expects to be paid for ferrying the departed soul across the sea to the immortal world and the gatekeeper at the gate of heaven needs payment to grant entrance into heaven.

Currency seems to have evolved from commodity money by 650–600 BC. In later periods, historical accounts see this evolving into a system of representative money. The initial use of representative money owes its origin to the lending system, which involved issuing some form of promissory note to the depositor. Subsequently, it functioned as a measurement of value in a transaction and depended on a parallel system for a guarantee of the value being exchanged, which changed to the gold standard in the modern era, before getting dislodged when the gold standard retired in different countries. It is now customary for the central bank in an economy to guarantee the value of a unit of currency on behalf of the state.

ACCOUNTING AND ACCOUNTABILITY: A BRIEF BACKGROUND

With this brief history and given that trade and commerce were established practices from prehistoric times, we can assume that for successful market transactions to take place, some form of recording business transactions was needed. The author hopes scholars will unearth some of these systems in due course of time and expand our scholarship. However, historical evidence of recording mechanisms is currently sparse, especially from the period before written records started. For example, we are aware that memorizing and reciting *sutras* kept the Vedas alive through generations before they were scripted on parchment to save the effort. In Indian tradition, *bahi-khata* accounting has been practised by merchants and business houses since the Vedic ages. In some cases, these accounting books have also been used to capture rich descriptions of the events and daily rituals in a king's court, going beyond numbers, commodities of exchange and financial settlements, providing us with an archaeological treasure from a bygone era (Nigam 1986). The details also reflect the responsibility of court accountants to cover transactions with details that were not limited to recording monetary matters alone. While this could be seen as a form of accountability, we can make an educated guess that accounting and accountability remained divergent, especially from modern Western perspective.

The Western form of accountability as it relates to accounting is confined to the concept of the obligations of an account writer, which originated in the lending practices of Europe, and in turn might have had its origins in the trading and lending practices of businessmen in the Middle East. This is distinct from accountability as understood in ethics, with the two remaining disconnected until recently. In contrast, the overarching Eastern view of accountability relates to moral responsibility and being accountable for intent, actions and outcomes, within which accountability in accounting includes recording the original events as a faithful narrative. Naturally, this does not square with the modern form of accountability businesses are expected to uphold. To me, accountability in accounting (in its current form) has been limited to being a mechanism of ensuring trustworthiness of accounting procedures and records, divorced from ethical aspects of the term, until recently, when critical theorists relied on the latter to argue that accounting should play a larger role in addressing social issues, including contributing to the quest of sustainability. This relates to the debate whether accounting, as a science and a practice, can be held accountable towards such expectations; but more of that later.

We now switch gears to learn a bit about sustainability and how it relates to accountability and, therefore, to accounting. Concerns regarding sustainability and ecological challenges serve as the background here, which has forced all branches of scholarship to revisit the fundamentals. Being a student of management and business, I have chosen to contribute to this evaluation and, in the process, explore concepts from economics, accounting, and business management, and their practical enactments. To decipher the entangled developments in these areas, we first examine how the sphere of human activities has become toxic for nature.

SUSTAINABILITY AND THE ANTHROPOCENE

The Western scientific approach with its over-reliance on reductionism and the deductive approach supported the Industrial Revolution and machine culture, where easily available fossil fuels contributed to economic growth while undermining our shared responsibility towards nature. Not surprisingly, the market culture helped us replace progress with affluence and accelerated human-induced imbalances in nature, impacting the biotic as well as the abiotic world that economic thinking chose to ignore. It is only in the last few decades that the scientific understanding of these human-induced challenges has become prominent. Here, the term 'scientific' relates to the advancing of knowledge through careful observation and experimentation, by employing scientific methods that are not biased to promote specific perspectives and interests but to advance collective knowledge based on facts and evidence. However, our contrasting worldviews are a cause for concern. For example:

- When we talk about challenges such as genocide, human and animal trafficking, war and political unrest, dwindling resources and overdependence on non-renewable sources of energy, how much of these can be related to economic interests? This can be a subject of serious research. Nevertheless, controlling economic resources and markets have been the most favoured weapon since the industrial age and this approach has shaped human affairs globally, bringing misfortune to a majority of the population and the natural world.
- With the human population all set to cross the staggering 8 billion mark, the space available for other species is shrinking drastically and proving to be a deterrent to the healthy growth of humans as well as other species. Could this be leading other species to become extinct much earlier than they would have otherwise? Humans, with their knowledge, have colonized the earth and its resources for their own selfish needs, avoiding accountability to Mother Nature and other species.
- Advancements in modern medicine has improved average life span of humans, not to mention improved probability of saving every normal and not-so-normal human foetus, baby, and child, and adult from life-threatening challenges, and we all are appreciative of it without realizing that, in the grand scheme of things, we are working against the order of nature. While medical advancements and human compassion are rightly justified in helping us prolong our lives, I am being politically incorrect in echoing the sentiments of other great minds who have cautioned that this and other types

- of technological advancements are too challenging for the natural order to be allowed to persist.
- Although critically assessing the contributions of human species towards Mother Earth has proved difficult for our combined knowledge base, to question impact of human activities on biosphere and to arrive at acceptable resolution is long overdue but premature, considering how different schools are problematizing the concerns and risks, including defining nature and everything else as 'external' to the domain of human sphere (defined variously as the anthroposphere and the Anthropocene).

Sustainability is a relatively new concept. Its definition even at its most basic level (at least to me) fails to shed anthropocentric overtones, as it is us humans who are the ones defining what sustainability is or should be. The human mindset is deeply committed to ranking and comparing, hailing one supreme over others—in terms of choices, decisions, social orders and behaviours. We live with the most basic abstraction of holding 'the' one as important or meaningful as against available others, which brings us to anthropocentric perspective, one that weighs sustainability concerns heavily in favour of humans. Years of training and scholarship have conditioned human minds to search for the 'best methods', the 'best findings', the 'best results', the 'best way', and so forth. Sustainability, as viewed from this anthropocentric perspective, unfolds as a challenge to human societies' desire to continue their dominance over the earth and its resources, reducing other species and resources to minority statuses, subservient to human interests. Even though this may be politically incorrect to say, the development agendas pursued by governments and societies across the world have (*mostly*) been following this philosophy, sweeping aside concerns of pursuing economic progress while targeting affluence levels. This reduces the question of planetary sustenance to a certain rate or methodology of development, embedding sustainability within development-related challenges, and shaping it to acquire a methodological perspective, as against questioning the overall philosophy of human progress and the corresponding detrimental impacts on the planet. At the same time, experts are yet to converge to shape and influence policy and governance frameworks so as to improve the collective approach to sustainability, so that instead of being 'collective' or focused action, it has evolved into a spectrum of overlapping ideas.

The Millennium Ecosystem Assessment (2005) has highlighted severe constraints on natural systems and resources such as arable land, stock of underground water, freshwater sources and other natural resources, not to mention the plight being faced by a significant population due to poverty, hunger, deprivation, human trafficking, and the list goes on. I cannot help but wonder why we are unable to acknowledge these as challenges of our own making and due to our collective outlook (especially those impacting sustainability), even when the signs are everywhere. Experts and scholars have never united to define sustainable development and sustainability to mean the same thing, and polarized views adhering to anthropocentric versus ecocentric philosophies have challenged different schools of scholarly thought. Leaving semantics aside, the basic question should be, 'Whose sustainability do we seek?' I do not think that Nature's benevolence favours the existence of one species over others or that she intends to benefit any one species particularly as a part of her plan. For example, the tension over meeting the present human needs within the boundaries of the providence of Nature and yet not trespassing on the sustenance capacity of the planet that rightfully belongs to a future generation also reflects a primal fear of losing human control over the earth at some future point of time. So do we need sustainability for the ecosystem in its entirety or is it the sustenance of the human race that concerns us more? The outlook is disturbing, but more so, due to our inability to frame the concerns and relate these to the boundary conditions:

- Although critically assessing the contributions of the human species to the well-being of Mother Earth has proved difficult despite our combined knowledge base, it relates to our inability to connect what we scientifically know and what we believe we should know. What are the things we need to know accurately (read scientifically) before we start working towards sustainability meaningfully?
- Another part of the equation needs us to work with the available signs—signs that do not offer a deterministic picture of the world in a way we are used to when planning interventions, but relay certain symptoms that we need to work with, without the knowledge

of how our efforts will finally shape up. The available knowledge base has allowed us to develop a few measures and frameworks that relate to regional and/or economic impacts. To relate these to the stocks and flow of natural resources, including how human interventions are reshaping these, and cataloguing everything to develop a taxonomy is a daunting task.

- We must realize that we risk understating the extent of our anthropocentric activities and the changes that these introduce to their ambient environment in its entirety, given our inability to examine the interactions at granular level. For example, our collective knowledge base has no way of ascertaining the carrying capacity of Mother Earth, nor a way to measure the depth of turmoil that the Anthropocene has been causing.
- Ecological sciences have added to our knowledge base how natural systems adjust for sustenance and counterbalance adverse impacts, which can make for natural delays before the impacts of larger changes become visible. However, as the human knowledge of causality is encapsulated within a time frame, the impacts of the Anthropocene that are delayed beyond a short window of time and further fractured due to the decomposition and diffusion of effects, results in a blurred view that our popular sciences are not suitably equipped to deal with.

Given these complexities, one thing is certain. The issue of sustainability has tested the limits of human knowledge and has forced theorists, scientists and researchers to evolve paradigms that can capture the problem with better clarity. It is forcing the scientific community to break away from the traditional forms of science that have been the bedrock of industrial developments. However, it would be challenging to get rid of our current habits. The prevailing scientific paradigms of deductive analogies help in isolating the anthroposphere from nature, which leads us to view anthropogenic interactions as neatly separable and accurately identifiable packages, obliterating the need to address the complexities in situ. By leveraging systemic perspectives in scientific studies of the built environment and the Anthropocene, we can better modulate our understanding regarding interactions with Nature. This is where scientific studies involving new areas of disorderliness could be of immense value—employing, for example, chaos theory and complexity theory, not to mention other analytical frameworks such as wicked problems, soft systems methodology and post-normal science.

In contrast to the anthropocentric worldview, ecocentric philosophy views coexistence of all natural systems as equally important, one influencing and interacting with the other, but no one gaining greater importance than the others. This approach calls for changing our views regarding business firms and their place in human societies. Ecocentrism accepts nature as the central force supporting the life system on earth, where everything existing within the biotic and abiotic worlds has its own reason and place. According to this philosophy, irrespective of how we view reality (from scientific perspective) or consider ourselves (the human species) intelligent and empowered than any other living species and non-living things, we should be working towards improving nature and its subsystems even when our knowledge of the intricate web of life systems is incomplete. In other words, ecocentrism intends to shape the collective intent and behaviour of the human systems towards nature and natural systems in a way where our worldview improves to hold nature to be the central force of ecosystems that it has evolved over millennia and continues to evolve, instead of centring the riches that nature offers for segregated benefits to humans.

One of the earliest forms of philosophical encapsulation of ecocentrism that I can refer relates to the Vedanta or the Upanishads. The Maha Upanishad, chapter vi, sloka 71 has the phrase वसुधैव कुटुम्बकम् (*vasudhaiva kutumbakam*) in Sanskrit, which translates to 'the world is a family' (Warrier 1953), where *vasudha* in Sanskrit translates to Mother Earth (as against the material earth), who uses her regenerative and sustenance capacities to maintain the *kutumbakam* or 'family', which in the fullest sense of the word not only relates to the physical aspects of the relationship that characterize members of a family, but also the caring, sharing and nurturing activities and connecting the earth and everything else, the entire microcosms of the biotic and abiotic worlds, and beyond. This unification is connected through '-*aiva*', which translates to 'indeed'. Although the usage of this phrase in the original sloka in the said Upanishad was to differentiate parochial or insular views from tolerant ones, its self-contained meaning aptly defines ecocentrism.

Having said that, as changes within the anthroposphere would anyway impact nature, intentionally or otherwise, and would contribute to changes in the ecological profile of a region, loss of species, land- and biodiversity-related changes, how would the ecocentric approach practically differ from business-as-usual thinking? This calls upon us to look at and learn from every discipline, including learning from aboriginal societies and how they have been living in harmony with nature for centuries. The idea is not to advocate living at a subsistence level, as discarding material happiness could be a *very* challenging option(!), but rather learning to live based on the principles of sustenance and harmony, for example, direct consumption of what nature provides, a cooperative form of existence, shared public goods that are equally cared for (as compared to the tragedy of the overgrazed commons that exemplify welfare economics) and greater adaptability to natural changes. Against this background, we move to the core of our interest—the economic rationale of human endeavours and how that squares with our accountability towards nature.

ACCOUNTING FOR ECONOMIC GAINS VERSUS SUSTAINABILITY: SHORT- AND LONG-TERM VIEWS

With the two contrasting premises from the previous sections, my aim is to take readers on a journey exploring how societies are learning to recognize and accept the implicit cost of development (read material progress) for the first time since the industrial revolution, but are yet to evolve a complete knowledge base. For example, if we are holding industrial progress at the core of our existence, we would need to tie sustainability to some form of industrial progress, maybe by achieving a certain rate of growth, working with specific industries that are hurting the proximate environment more than the others or benchmarking industrial performance. Crucial to this effort would be to also question whether holding on to this worldview might be causing more harm than good—but who is going to decide that? Let us review two examples: one from the tobacco industry and the other from the large dam projects. There could be other examples as well, to substantiate or refute some or the other argument presented here; however, the ones selected offer situations that have unfolded globally and reflect economic exploitation of resources (in the first case) and the expansion of the technocratic approach leading to large-scale impacts (in the second case), causing social and environmental challenges of universal nature.

Case 1

While tobacco has been a major recreational drug for aeons, its commercial exploitation in the form of rolled tobacco or cigarettes has been very recent—say, over the last 150 years or so, although widespread adoption of this form of smoking tobacco is a 20th-century phenomenon, when the per capita consumption of tobacco peaked in the adult population in the developed world, whereas in the developing world cultures, it was adopted mainly by the male population initially (Meinking 2014). Despite the cultural and anthropological interest that it carries, how this sociocultural phenomenon impacted the prevailing social norms of the time is a subject outside the scope of this book. However, German scientists were the first to link smoking tobacco and lung cancer in the earlier part of the 20th century; by the end of the Second World War, the risks were known and identified in scientific literature (Proctor 2000). Yet it was only in the early 21st century or at the fag end of the 20th century that different countries finally agreed to label cigarettes and packaged tobacco products to warn of their cancer-causing potential and globally advanced to implement tobacco-free spaces. So, while science could establish causality, it took close to half a century to integrate this into our collective knowledge base. The important question to ask is why this delay. Maybe because the benefits (for businesses) of having tobacco socially acceptable as a recreational drug remained disconnected from the impact (for society). Let us hold on to that thought while we review the second case.

Case 2

This is related to the commissioning of large dams as part of infrastructure projects in different countries. Here the discussion is restricted to the standard industry definition of dams that are higher than 15 metres, or roughly a four-storey building. There are other parameters attached to defining 'large dams', but this discussion is not about the technical aspects. It is about our fascination with building larger and grander versions of the engineering marvels that we love to hold up as human achievements. Estimates vary regarding the number of large dams that have been built around the world, but the number might be anywhere near 60,000 dams. China alone has 23,000 of these and the United States (US) another 9,200. Globally, these dams have flooded reservoirs of the size of the state of California, more than 400,000 square kilometres, which is 0.3 per cent of the world's land, not to mention the loss of fertile river beds, diversified ecosystems and wetlands. As per one estimate, roughly 40 to 80 million people around the world have been affected, the majority in India and China, including lives lost due to flooding and dam collapse.[2] Dams help us with cheaper hydroelectricity and provide water to irrigation systems for agriculture, in addition to other benefits that have been cited as reasons to build these dams. While the increasing acceptability of environmental impact assessment (EIA) studies before undertaking large projects is a matter settled in law, literature on this subject is inconsiderable.[3] Interested readers can refer to the book *Silenced Rivers: The Ecology and Politics of Large Dams* by Patrick McCully (2001, originally published in 1996), which undertakes socio-politico-economic analyses of large dams. From our perspective, the legal emphasis on EIA has to an extent allowed multiple stakeholders in a project of large magnitude to voice their deliberations, beyond the confined zones of 'returns' and 'costs', and allowed decisions to be more 'inclusive' in nature. However, one reason for the slow proliferation of scientific studies on the downside of dams could be that these are not felt in the same area of science from where the solutions have been created in the first place. These are technological solutions but are causing ecological, environmental and social challenges.

We can question whether dams are the only source of cost-effective electricity generation or whether to allow policymakers to understate the social costs (be it treating cancer from smoking tobaccos or the cost of rehabilitation of people displaced by megaprojects) that far exceeds contributed welfare costs (read taxes by the concerned industries). If

[2] Statistics sourced from the International Rivers website (2016).
[3] Google Scholar listed only 647 results for a search with keyword 'damming' and 14,100 with 'dams' in the title (with no date range), whereas the keyword 'management' in the title generated nearly 1.2 million results.

the answer to the above is yes and yes, we can collectively agree that human societies are yet to learn to be truly inclusive and consensual in approach, transcending circles of power and controlling interests, to bargain for the overall welfare of the society. If this is the case, what are the chances that human societies will discard their anthropocentric thinking in favour of something subtler and more inclusive, like sustainability? Moreover, what would be the role of industries and businesses in supporting the 'inclusive' approach that is yet to emerge from theoretical shadows? We can also generalize that the most prominent voice in a social or political context mostly remains disconnected from the fragmented or distributed consequences, which in turn takes a longer time to make a mark.

As the fundamental premise of the ecocentric approach is the simultaneous coexistence of all natural systems, no vantage point is to be expected by privileging one plan or system over others. Accordingly, from an ecological perspective, any intervention planned within a state of equilibrium—as against another, howsoever rational—can only be a learning process. What would be the role of business enterprises within this changing paradigm, one might ask, as the diverse viewpoints hardly converge to explain how industrial development can lead to thinking about sustainability. This is a concern that goes beyond the moral philosophy of sustainability, as any successful approach has to be able to guide scientific decision-making and influence the behaviour of firms. As our collective conscience has remained grounded in human achievements for most part of the human history, polyvocal considerations could lead to fractured reactions towards social institutions and businesses as well. Businesses, in contrast, may choose to react to only prominent voices and resort to superficial responses to handle the more scathing tones, just enough to be perceived on the right side of the divide. One of the significant advances in this area is the advent of an industrial ecology and the perspectives it offers, where the collective behaviour of firms and how the built environment relates to its surroundings are being considered as equally important areas. Interested readers can refer to Gradel and Allenby (2003) to explore industrial ecology and related insights. However, this is not far from exploring the mechanical view of things, limited to uncovering interactions between entities, instead of experimenting with different forms of enacting sustainability and developing systemic views about these.

CONTEMPORARY POSITION OF ACCOUNTING SCIENCES, ACCOUNTABILITY AND SUSTAINABILITY CHALLENGES

Since the advent of joint stock companies, accounting and related methodologies have played an important role in shaping the beliefs and actions of the management, stakeholders (earlier shareholders), markets and society, with respect to how they would view the performances of firms and evaluate investment decisions. Accordingly, it is natural for firms and market to rely on accounting as a language of choice to decipher if a firm is responsive towards changing realities. However, even when accounting was the language of choice to relate to organizational performance, it was restricted to describing the financial affairs of firms, and not particularly suited to resolve concerns such as interpreting the impacts of businesses on the environment and on society, due to the lack of certain structural elements, such as the incompleteness of contracts, missing legal statutes to enforce responsibilities, inability to convert environmental aspects into monetary equivalents and the lack of standardized interpretations. While assigning monetary implications to these aspects is one part of the problem, developing objective interpretations for reporting and decision-making is another. A physical exchange with proximate environmental impact generates what an EMS defined as an 'aspect'. An aspect is thus the physical impact that a transaction generates and/or disposes of—for example, waste, emissions and other materials generated from the process. Environmental impact from the perspective of EMS is how the aspect changes the environment. In case of accounting, environmental impact is defined as the economic considerations to counter, nullify, or minimize the negative outcomes of aspects, as in some cases the changes caused by aspects could be irreversible (like extraction activities), and would need rehabilitation, which is also considered as externalized liabilities (liabilities not acknowledged by the firms).

Given these concerns, it is difficult to figure out whether there is a deliberate attempt on the part of the firms and prevailing accounting and reporting practices to avoid incorporating the environmental impacts of firms in their reports, or whether this is the logical fallout of the inherent qualities of the accounting sciences, which as a language of business performance cannot deal with such situations. Equally relevant here is the fact that if the accounting methodologies fail to provide constructs that firms can rely on for evaluating and tracking environmental risks and if they are unable to warn societies about the impending loss of natural capital that is getting traded off for economic benefits, it ultimately translates into a shortcoming of accounting as a discipline. A part of the responsibility for this failure will have to be borne collectively by the accounting scholars, business managers and practitioners, whose efforts have fallen short for firms to contribute to environment and societal concerns. So the need of the hour is to not let organizations shortchange sustainability requirements by employing reporting solutions which can hardly be an effective method to improve material usage, energy intensity and waste-generation parameters of products and services in a transparent manner, and instead to develop a conducive environment to experiment with innovative methods and tools that have emerged within the discipline and to critically evaluate the available choices.

Interestingly, accounting for environment has been in the lexicon of research and literature on the environment, where scholars have experimented with different ways to express how a firm is performing environmentally, rather unsuccessfully so far. Although the firm–environment exchange is inherent to the existence of firms, the study of these interactions under the aegis of industrial ecology is nascent and has been gaining momentum only in recent years under the umbrella of social and environmental accounting (SEA). These developments have highlighted concerns that businesses are expected to deal with, such as lack of accountability and concerns over sustainability. As a practitioner and researcher from this field, I am inclined to ideate and explore new ideas, for example, to evaluate the role of the accounting sciences beyond the rudimentary level that business firms rely on and to discover how they can meaningfully incorporate accountability, meeting the demand from critical theorists for firms to participate in social challenges beyond what is actively pursued in the research corridors.

Another aspect of this debate concerns the inherent reality that consumers of accounting information depend on, which includes the contextuality of the information that the accounting artefacts disseminate, where the generalizability of information is based on its adherence to the accounting standards. So it can be argued that accounting sciences are under no compulsion whatsoever, within the normative and positivist accounting theories, to capture the interactions of a firm beyond the economic implications of doing business.

At the same time, holding accounting practices responsible for the lapses (of not going beyond the economic implications of doing business) and the short-sightedness (to overlook long-term impacts of business conduct) that it is designed to preserve, citing ethical reasons or otherwise, might lead to abrupt changes. If pushed to it, the character of a system that is designed to function and deliver information by stripping out details becomes argumentative and counterproductive—for example, in defining the social and environmental impacts of doing business. Thus, some of the questions that we would be exploring and experimenting with on the journey through this text are:

- Why are the accounting sciences unable to reflect the firm–environment exchanges accurately?
- How could we use the accounting language to capture these interactions and what kind of interpretations might this lead to?
- What would be the new methodologies needed to capture information and improve the decision-making functions of firms?
- How does the changing vocabulary of externalities, full costs and natural wealth challenge the core accounting concepts such as costs, certainty, assets and liabilities?
- What kind of methodological improvements in accounting can help firms connect with sustainability?
- What would be the direction of such evolution?

Last but not least, an important objective of this manuscript is to contribute to the ontology of the accounting sciences, which have been waiting for a long time to evolve beyond the utilitarian values that they have traditionally been confined to. So while connecting accounting with accountability is a key step, accounting is also examined in terms of the core value system it has traditionally been assumed but hardly ever explored beyond its calculative capabilities. To achieve these objectives, the manuscript has been arranged in four sections. Each of the sections is introduced here, with a brief description of their contents.

Section I. Accounting and Accountability: Traditional Paradigm

The first section captures the traditional paradigm within which accounting functions. It assumes that readers have an understanding of the basic concepts of accounting, reporting and accountability, and the boundaries within which accounting traditionally operates. As is well known, accounting is also considered a language of business, being used to measure, interpret and record business transactions on the input side. On the output side, accounting supports internal and external reporting requirements, where the recorded information is reproduced in different forms to meet the needs of internal and external stakeholders. Dissemination of information to external stakeholders also acts as an information signal to the market, as Ball and Brown (1968) could empirically prove. Accordingly, the accuracy of reported information and the interpretation it may have brings in the role of accountability, beyond regulatory bounds. To appreciate this relationship, accountability has been explored through the lenses of economic, organizational and ethical theories. Even though the relationship between accounting and accountability is believed to be a persistent one, the findings suggest it is more of a rhetoric. In other words, accounting (or whatever theorization has been achieved so far in this area) is yet to embrace accountability as a universal value system. This is where we see the argument for firms to disclose qualitative information along with the quantitative ones as a bid for increased accountability. This also brings in the interpretive nature of disseminated information, and its ability to influence the users of information, even when the information is assured, and expected to be neutral in nature.

Next, accounting and its contemporary concerns are explored through the two predominant forms of accounting constructs—financial accounting and management (managerial) accounting.

Financial accounting is explored in terms of the underlying economic views and how accounting standardization contrasts with concepts that have been tested through time (the income view versus the balance-sheet view). On the other hand, management accounting is explored from the perspective of the business decisions and related information needs that reinforce calculative perspectives and economic paradigms. Findings from research corroborate the position that even managerial accounting has struggled to transcend the adoption of a purely economic view of business, even when it experiments with latest decision-making techniques. Given these limitations, the paramount question is whether accounting is equipped to participate in addressing the contemporary challenges that firms are facing, including its role in supporting the sustainability beyond economic sustenance, and to help businesses improve their social narratives to match that of a responsible citizen.

Section II. The Environment and Accounting Theories: Contemporary Advances

The second section examines the state-of-the-art of the problems and issues that the businesses are facing, including increased scrutiny by stakeholders, risk of regulation, multiplicity of voluntary frameworks to choose from, and a diversity of approaches in addressing some of these. This includes the nature of firm–environment interactions and how it relates to the wider disconnect between the anthroposphere and the sustenance of Mother Earth. The disconnect underlines economic sustenance and the inability of firms to connect to environmental concerns beyond altruistic interest. This results in exploring the changing role of businesses to fulfil the needs of modern society and the ecological challenges that they are contributing to, not to mention the plethora of overlapping reviews, solutions, frameworks and expert opinions advising how to handle some of these challenges that are not always well-aligned to the stated objectives. The review also relates to the role of contemporary accounting practices to participate in some of these concerns and highlights vexed issues such as the ownership of waste, legal challenges in ascertaining liabilities associated with public waste, and systemic perspectives of externalities.

Against this background, contemporary developments in green(ing) accounting are explored, including some of the recent theories and developments in environmental management accounting (EMA) within the field of SEA theories. Methodological developments in EMA are explored in this section through a review of different areas of research and advances in corporate environmental sustainability frameworks, which include emissions and carbon accounting, sustainability reporting and corporate sustainability reporting frameworks. The study of these areas is relevant to the ongoing developments in the field, as scholars and practitioners are working towards developing techniques—such as environmental life-cycle tools and materials flow analysis—to help firms ameliorate the environmental perils that they contribute to. Accordingly, waste and emissions have been discussed in terms of measurement challenges as well as the lack of institutional approaches that firms might be looking for. To summarize the findings in this area, these advances seem fragmented at present and do not offer any overarching solution that accounting sciences can rely on. However, one view remains dominant. To handle these issues better, in addition to methodological solutions and environmentally enhanced technologies, firms would also have to invest in improving their products, services and processes as well.

Section III. Environmental Accounting: A Dimensional View of Accounting

This segment is all about finding the right path to address the challenges to the accounting sciences and connecting critical views with the normative theory of accounting, so as to enable them to simultaneously encapsulate the environmental and economic performance of firms. Other than the externalities, environmental accounting is expected to improve the accountability of industries towards the aspects they produce, something that cannot be settled amicably by resorting to a half-hearted approach. While it is easier to use quantitative methods to generate information that is blind to the business intricacies and temporality, this is definitely a risk. Even though the prevailing theories have been accommodating new information demands by extending themselves, a more pragmatic approach to dealing with these increasing pressures would be to think of an overarching solution. This is where the unidimensional nature of the prevailing accounting construct and its overwhelming support of the economic existence of firms alone forces us to rethink the problem. Accordingly, it is postulated that no amount of change within the existing framework(s) would cater to environmental needs, leaving care for the environment leaving it disconnected from the economic performance of firms.

This is where environmental accounting is introduced as a paradigm shift and explored in successive chapters of this section that rely on published case studies to model how the conventional views of accounting and contemporary developments are still falling short of the evolution needed to cover the requirements holistically. This results in exploring valuation or monetization methods in detail and improve the narrative around externalized liabilities based on repository of real transactions. Leveraging experimental costing techniques, advanced cost models are developed to formulate a 'monetization token' that firms can use to translate their environmental performance in the ordinary course of business. Based on the stakeholders' needs of improved information on environmental performance and decision-making, environmental accounting could operate independently—as a new framework of accounting—and link to financial and cost accounting through underlying transactions. This accounting framework could exist in parallel but remain conjoined to the existing accounting frameworks, which would need new sets of rules for its institutionalization and effective handling of auditing and reporting needs. In this process, different costing and management accounting techniques are experimented with, to quantify and monetize the environmental impacts of wasted resources and materials, all this and more that fits into the construct of environmental accounting.

Section IV. Accounting Sciences and Sustainability Theories: Managerial Implications and Recent Advances

The last section is a compendium of the contemporary developments in this field and the managerial implications of these studies, with the last chapter in particular detailing some of the recent advances on sustainability and accounting. In terms of managerial implications, carbon accounting is explored in terms of emissions (physical) and carbon (normative) accounting, designed to handle carbon trade-related transactions and to help develop an integrated waste flow. To address the greening needs of firms, environmental management systems (EMS) and green information needs are explored, revealing how firms would have to depend on a mechanism to capture, calculate and record data and information and help them take corrective action. This is where the environmental paradigm of accounting is found to be complementary to the EMS. This discussion is followed by an overview of some of the recent advances in sustainability reporting, and their institutionalization has been examined. The last chapter studies advances in the ecological sciences and their views on sustainability, while considering how accounting can evolve to safeguard ecological interests, followed by a discussion of the need for a multi-dimensional accounting construct to handle complex challenges such as human-induced pollution in space which our technological advancements are adding to. This relates to the future potential of developing a multi-dimensional accounting construct that adds to the longevity of the accounting sciences. Future research can advance through experiments leveraging the concepts proposed here and improving upon ideas such as ecological accounting, green information systems, green projects and strategies, and ecologically enhanced business models.

The present chapter has served as an introduction to the text, laying the foundations and overview of the focal areas covered in the manuscript. While the book is in four sections, dependencies across sections are minimal, which allows readers to seamlessly move from one to the other. Readers interested in advanced concepts can directly jump to later sections and explore these. I, the author of this text, take full responsibility for the content, any inadvertent omissions and mistakes as well as incompleteness due to inaccessibility of certain research. This is only a work in progress, an on-going dialogue, inviting all to participate.

Accounting and Accountability: The Traditional Paradigm

I

Organizational Theories and Accountability

INTRODUCTION

Literature theorizing the rationale for the existence of firms reveals multiple perspectives and include economic, managerial and institutional views of firms. In this chapter, some of the relevant theories are explored to define the accountability of firms towards society in general, and to examine whether the concepts can be extended to sustainability in particular. Theorization on accountability is vital to relate the way in which it is practised in firms and the way organizational behaviour is shaped in response to disruptions that are commonplace in the competitive arena today, not to mention examine why it has become a concern of late. Organizational accountability being referred to in here corresponds to the institutional accountability and is different from how it has been defined within the accounting scholarship.

Accountability from the accounting perspective is relevant to explore the flow of information that is guided by the regulatory frameworks, and to examine voluntary disclosure practices; including exploring whether that overrides certain other aspects of accountability or distorts it in places. In contemporary research. Accountability has been explored from the point of view of traditional as well as critical theories and its study is gaining prominence to substantiate increased *activism* of firms within the domain of sustainability. This chapter also investigates how accounting and accountability overlap and the implicit and explicit effects of their connection rely on each other to handle ethical stances in a postmodern industrial world, where firms are expected to reassess their accountability to society, nature, future generations, stakeholders and regulatory institutions—in short, all entities that influence or are influenced by the firms' actions and decisions (read behaviour) across different time horizons.

ORGANIZATIONAL THEORIES OF THE FIRM

We begin by exploring the organizational theories on the existence of firms and how these theories connect with the domain of accountability. The evolution of business firms and separation of ownership from owners or suppliers of capital has been a major breakthrough in economic progress and in developing the capital market. This has also fuelled the entrepreneurial spirit by enabling access to funds that the new projects would need and rapid replication of micro-entities fulfilling societal aspirations. However, the fundamental question why firms need to exist, remains. This has been answered within the theory of firm, that reflects boundary considerations of the firm, their structural arrangements, heterogeneity, and evidence from different theories seems to have answered a few of those in parts. But, before advancing any further, we might ask why we need to delve into this matter.

In recent times, the justification of firms being a social institution has been questioned (Kolk, Levy and Pinkse 2008). By legitimizing economic entities as artificial 'persons', the outlook shifts the perspective and rationale for their existence beyond the creation of economic value. This includes questioning how firms, being 'responsible citizens', can contribute to larger, challenging social issues. If firms are viewed as social institutions more than economic entities, the question shifts to organizational choices and ethics, beyond the prevailing theories on firms. Accordingly, our intent in examining economic and management theories rationalizing the existence of firms is to find a connect with the ethical theories. From economic theories, we examine Coase's transaction cost theory and Williamson's theory of asset specificity, whereas among the management theories on firm, the theory of profit and sales maximization as well as optimization theory have been discussed. The behavioural theory on managers has also been explored to argue that organizational well-being driven through the maximization of profit or sales is a myth in the long run and a counterproductive strategy for firms to pursue. From the school of ethical theory, we have relied on ethical approaches and the theory of ethical rights to examine the accountability of firms. While there are other theories worth investigating, the discussion has been limited to the ones that correspond to the nature of this investigation. Readers interested in exploring these and other theories further can refer to the standard texts on the subject.

Theory of profit and sales maximization

The theory of profit maximization was developed as a part of classical economics by Walras and Marshall in the 19th century and remained unchallenged till 1920. It explains the existence of firms being necessitated to maximize returns for the shareholders in the short and the long term. Theories based on the goal of profit maximization hold this to be the single most important reason for the existence of the firm, which corresponds to the shareholder theory in management literature, where management holds a fiduciary duty towards the shareholders to maximize returns on the capital invested (by earning higher profits, paying dividends and improving stock prices). Based on the concept of improving marginal profit, the theory of profit maximization defines the optimal level of operation as the one at which the incremental profits match incremental costs. Skipping a mathematical analysis of this view, which interested readers can refer to from the standard texts on the subject, our discussion is related to the basic tenets of the theory that hold firms responsible for generating better returns for their shareholders. In this process, the time value of money plays a dominant role that helps firms discount future profits in favour of higher current profits and aligns decision-making accordingly. Even though the major determinants of firms' decisions on pricing and quantity of products and services are dependent on the market situation and beyond the control of the firms per se, the economic theories of profit maximization view firms to drive value maximization.

The deficiencies of the theory of profit maximization were addressed by Baumol's theory of sales maximization. The central argument of Baumol's theory is based on the uncertainty of the underlying market conditions and the competitive forces within which the firm operates, which kind of makes profit a variable outside the locus of management control. This theory postulates that firms (and their management) aim for direct and tangible goals that the firm can control and improve their sales or top-line performance. Here, higher sales translate to higher profits, so long as the factors influencing growth of sales remain within the control of the organization. However, optimizing sales would mean not to cross the boundary of performance post which the incremental sales would lead to a higher increase in costs as against profits. The other part of this postulate relates to the justification for management to sacrifice short-term profit in favour of higher short-term revenue—by, for example, promoting a product mix with higher value—to improving the top-line performance. Unfortunately, Baumol's theory has not always found empirical validation, being contradicted at times and sometimes supported in specific cases. Moreover, variations such as advertising and sales promotions have been studied in relation to short-term revenue maximization, but failed to generate uniform conclusion. Accordingly, the generalization of economic theories to explain the existence of firms is about economic sustenance of firms, which in turn means more value for the suppliers of capital.

Transaction cost theory to explain the existence of firms

As opposed to the theory of profit/sales maximization from classical economics, an empirical theory explaining the existence of firms in neoclassical economics was proposed by Coase (1937), where the existence of firms has been deliberated upon through the lens of transaction cost theory. Even though transaction cost theory was improved significantly later to explain contracting and negotiations needs of economic agents, what it fundamentally relates to is the collective ability of firms to lower the overall transaction costs, compared to a situation where all activities are conducted through individual market exchanges (e.g., contracting for human skills by negotiating for individual contracts in the open market). What it means for a firm is that it exists to internalize part of contracts within its boundaries and economize on the cost of contracting. For example, by entering into contracts with employees for their services or for them to execute certain jobs, firms bring these contracts within their own operational boundary, thereby improving efficiency for the whole economy and lowering the overall cost of contracting.

Classical economics do not follow economic activities in such a way as to offer a narrative on how the factors of production combine to operate within a given market, whereas transaction cost theory offers a more realistic explanation as to how firms exist to lower and stabilize transaction costs. In this case, firms as economic agents encapsulate a number of contracting areas—contracting, for example, with employees or with other economic and contracting entities that would otherwise need to be in place and be driven by the market in order to carry out production activities. This lowers the overall cost and drives efficiency through the reduction of repeatable contracts, which the firm absorbs as a part of its operation and factors in the associated costs. This results in lowering of costs to regulatory bodies and policy administration where governance and regulatory bodies have to deal with only a limited set of transactions. A consequence of this argument is that we would expect the size of firms to keep on growing proportionately to improve their efficiency further and drive costs lower, which can collectively drive efficiency within an economy.

Coase related the efficiency gains to the size of the firm to postulate that the size of a firm could grow (in terms of the costs involved) until the incremental cost of entering into a new transaction (contract) matched the market cost. For example, firms would prefer to keep their recruitment and resourcing services internalized, so long as the cost of contracting talent through outsourced services remains higher. Although, this view has subsequently been studied and explored by other economists as well, the point of this being a simplified one remains an understatement, assuming—for example the cost of intra-firm contracts within an economy are lower than the inter-firm contracts within an economy. At the same time, this explains improving the bargaining power of firms to lower intra-firm costs, which in turn would explain why having employees as a part of the workforce results in lowering costs as compared to contracting options, or for that matter, resorting to marketing options for niche skills. This differs from modern contract theory, also known as the Grossman–Hart–Moore theory, which was developed based on an incomplete contracting paradigm that relates to the property rights approach and argues for the perennial incompleteness of contracts; consequently, this compels firms to try and cover every possible contingency and boundary condition in their contracting. The fallout of this directly relates to the risk elements that a firm needs to be able to comprehend and cover as a part of its contingent liabilities in financial reporting, which is however limited by the intricacies of contracting and the inability of any accounting expert to uncover and provide for every possible financial ambiguity.

Williamson improved upon the transaction cost theory of contracts by bringing in the theory of asset specificity, which can be explained in terms of the ownership and contracting boundaries of assets that a firm can use over time. According to this, the property rights of a firm would depend on the specificity of contracts and the sharing of pay-offs with suppliers. If contracting opportunities could cover the ownership of assets, firms would be able to drive production more efficiently by the sustained usage of assets instead of worrying about sharing the usage of assets with suppliers and thereby entering into frequent and uncertain contracts. This gives rise to certain hold-ups when the contracting parties fail to honour the contract due to its incompleteness. Labour unrest and organized labour negotiations are typical where an optimality satisfying both the contracting parties has not been envisioned before execution of the contract, and they result in an increase in the overall transaction costs. Although there have been solutions suggested, such as complex contracts and the removal of externalities (including information symmetry) from contracts, hold-up problems are still a common recurrence. Vertical integration (like mergers) have also been viewed as a possible solution (Whyte 1994) to remove the asset specificity towards shared ownership of assets. Vertical integration (such as in mergers) has also been viewed as a possible solution (Whyte, 1994) to remove the asset specificity towards sharing the ownership of assets. However, counterviews to this exist as well that argue that asset specificity supports non-integration (Kvaløy 2007; Ruzzier 2009).

Alternative and behavioural schools of thinking

Extending the theory of profit maximization to support long-term sustainability poses concerns, as a short-term focus on excessive profits can conflict with customers' interests or compromise the quality of products and services. This brings us to Williamson's model of the maximization of management utility proposed in 1964. Williamson's model effectively connects to contract theory, acknowledging that managements' goals may not be in line with those of the shareholders. It also squares with the agency theory of management, where the management's role is that of an agent of the owners. To align the two seemingly divergent views, Williamson's model proposes to improve the utility value of management, so that the focus of management aligns with organizational goals. This is an interesting theory as it explains the role of management incentives to act as enablers that entice the management to achieve or augment organizational wealth. The utility function of management that Williamson proposes is a function of number of direct reports, managerial slack absorbed as a cost, and a discretionary power of investment. Incidentally, this representation is based on the hierarchical structure of management generally followed in organizations, and in rewarding managers with better perks like discretionary investments that the managers can 'control' as pet project and feel 'connected'. Importantly, maximization of managers' utility model connects managerial intent and improved overall profitability as a function of the management's own existence, thereby connecting agency theory to the contract theory.

This brings us to Marris' theory of balanced growth (proposed in 1971), which is based on the divergence of ownership and control and seeks a reconciliation between the two. While profit and shareholders' values are the core interests of the owners of a firm per the profit and sales maximization theories, the management's utility value (in terms of salary, power, investments, and so on) is the foundation of Williamsons' model. Unlike Williamson's model, Marris' model holds a common element of interest to both the factions and also refers to the size of the organization. Marris' idea of balanced growth takes cognizance of the rate of growth of demand for the firm's products and of the growth of its capital supply. The argument is that the size of a firm can be based on different indicators such as revenue, capital, output or market share, without taking a definite view of one being more important than the others. Moreover, for balanced growth, this theory holds, it is not the absolute growth of the firm but the rate of growth that managers should attempt to maximize.

The models discussed above do not propose any upper limit for improvement, so managers can improve their utility infinitely, a supposition which was challenged by H. A. Simon (1956), who proposed the idea of 'satisficing' (a portmanteau of 'satisfying' and 'sufficing'), a bounded rationality, as the practical approach that is used by managers for decision-making. Simon brought insights from behavioural theory to management and challenged the basic assumptions on decision-making, replacing the idea of maximization (from the rational choice theory) with 'satisficing'. Satisficing is a cognitive heuristic for decision-making, where the acceptability of an alternative is not governed by rationality but by the constraints of available experience, expertise and prevailing conditions. If maximization can be considered as 'optimized' in a given decision-making scenario, satisficing can be equated with 'constrained optimization'. This caps the maximization approach with certain constraints emanating from managing a real-life situation where incomplete information would push for a near-sighted decision instead of pursuing a rational choice. Cyert and March (1963) go a step further to conduct an in-depth study of large enterprises where the management is separate from the ownership, reflecting that the goals of the participating entities in relation to the firm can be in conflict, and agreeing that a firm would always pursue what they referred to as common goals and that these could include improvements in production, inventory, sales, market share and profit goals. In other words, instead of maximizing the overall objective, sub-optimization or an attainable level of goal can be the aspiration, dependent upon a coalition of resources.

Economic theories and the accountability of firms

Economics, due to the very nature of the philosophy that it is based upon, strips out the nuances around the activities of micro-entities to compute the mega-trends of an economy, and as a result, side-steps the task of investigating the individualistic nature of business entities as well as their conduct. So while entrepreneurs might not have been running businesses with the mindset of 'it is all about profit' and could be more concerned with their livelihood, doing a service to the community, innovation, and so forth, the framework of the economic theories truncates these individual traits to focus on the organizational performance that a market and investors can objectively assess. This has resulted in the institution of an objective measurement of performance for firms through certain variables that derive a 'certain' value for a firm. Going by the above and considering how economic theorization on the existence of firms has shaped the business world, probably the questions that need to be asked are:

- Where does accountability figure in these existential theories of firms?
- Is the nature of accountability of managers towards owners any different from firms' accountability to shareholders and markets?
- Besides profit and valuation, can firms be held accountable for ethical issues, notwithstanding the legal and regulatory compliances that firms anyway need to be accountable to?

I gather from these theories that once the ownership of firms was divorced from the responsibilities of managing the firms and once the economics rationalized the management's utility (or success) in terms of their ability to continue the momentum and direction that entrepreneurs would have wished for it, the corporate character of the firm became synonymous with economic success, and other aspects of corporate behaviour—such as accountability, ethics and responsible corporate behaviour—became subfunctions of corporate governance. Corporate governance on the other hand relates to the available legal frameworks and narrowly defines how corporations should be controlled and directed, mostly in legal terms. Accordingly, theoretical advancements in examining corporate character remain confined to legal compliance on the one hand and the accountability of management to ensure sustenance, albeit in solely economic terms, on the other. Friedman (1970) echoed the perspective by pronouncing 'the sole purpose of a business is to generate profits for its shareholders'. From the perspective of institutional accountability, this is narrow and petrifying, especially considering the (unstated) faith societies have reposed in businesses and industries by entrusting them to deal with scarce resources, without holding them accountable towards their judicious use and after-effects. In contrast, the stakeholders' theory of Freeman (1984) expands the responsibility of management towards other groups of entities that impact or are impacted by the actions and decisions of management, but without offering a normative or critical view of how the newly assumed responsibility would be any different from the economic welfare that is the prime force of management or whether mere cognizance of the responsibility suffices for its delivery, much like Indian politicians for whom winning an election seems to be equivalent to discharging the promises that they have made during election rallies, without actually delivering any of the promises! This view is short-sighted and lacks the drive to infuse corporate accountability into firms when faced with competing decision choices (maximize profits, growth, employee engagement, corporate excellence, and so on), but more on that later.

The challenging ramification economists have faced from the trend of profit maximization relates to the uneven distribution of wealth that welfare economists theorize as a social welfare function rendered less than optimal due to market inefficacies and failures (spillage). Welfare economists have postulated that the loss of the welfare function due to market inefficiencies can be countered through a regulatory mechanism, such as a tax structure that can result in redistribution of goods and services, thereby improving the welfare function. So maximizing the social welfare function, from the point of view of welfare economics, relates to the equitability that the free-market system is expected to provide as a self-regulating mechanism (the invisible hands), whereas regulatory constructs handle spillage and reconnect these to the economic cycle. In essence, welfare economics transfer concerns from the input side of things (market imperfections) to the output side of things (redistribution through instruments of the command-and-control regime), unassumingly relying on the invisible hands of the market to alleviate all inaccuracies of the system, replacing ethics and quality of life with welfare through material prosperity. This, as I see it, is not a task for economics to deal with, but the fundamental design of the subject that blocks economic theories to connect to the accountability of firms in general and to societal and environmental challenges in particular. To explain these disconnects further, the next section explores the limitations to expanding the accountability of firms beyond their commercial performance, before bringing in inputs from the ethical theories to expand on organizational accountability.

ACCOUNTING AND ACCOUNTABILITY: IMPLICIT AND EXPLICIT OVERLAPS

Although from a common-sense perspective, accounting can be situated within the domain of accountability, accountability in its present form came into use much later. Labardin and Nikitin (2009) and Baladouni (1984) trace the etymological origin of 'accounting' or 'accountancy' to the vulgar Latin *computare*, which means 'to reckon'. The base of *computare* (to calculate) is *putare*, which means 'to prune, to purify, to correct' an account, and hence 'to count or calculate' as well as 'to think'. On the other hand, accountability stems from the late Latin *accomptare* (to account), a prefixed form of *computare*. While the words share the same origin, their usage refers to different aspects of businesses, as these two words gained prominence to apply to two different areas of accounting. One relates to the process (accounting) and the other to the responsibility of the account-giver (accountability), and the existence of one is held meaningless without the other.

Accounting practice in firms and its role in corporate behaviour (to the extent this has been theorized) have an undeniable overlap with firms' accountability regarding financial matters to shareholders, creditors, lenders and even the market and have been considered as sacrosanct, evangelized and regulated so that firms release and report financial data and information in a transparent manner. Accordingly, the accountability of the firms is inherent to the process of deriving financial information, recording and disseminating it. To ensure the due diligence within the accounting process that accountability demands, the practice of assuring financial affairs has been incorporated. This has been institutionalized to uphold the verifiability and neutrality of accounting information and to qualify accounting practices within firms, including reporting and qualifying accounting policies and financial matters that are, or can be, of material importance. By bringing in these conditions, the enactment of organizational accountability with respect to the financial affairs of the firm is ensured. In addition, accountability of information is assured by disseminating information with contextual specificity to help stakeholders take informed financial decisions.

To this end, accounting standards help accountants in capturing and generating information in accordance with the norms. Accounting standard as per the generally accepted accounting principles (GAAPs), including adoption of the accounting standards under International Financial Reporting Standards (IFRS), is based on the concept of stewardship. This includes holding directors responsible for accurately reporting financial performance of firms. Here, the role of accounting practitioners cannot be undermined, as they are the principal architects who define organizational performance in a manner prescribed in the accounting standards and as per the regulatory norms. Here, accountability does not mean simply conformity but upholding the sanctity of due process and reporting financial affairs without any bias, even though accounting has a lot to do with conforming to the accounting standards as accounting practitioners sift through the transactional data and translate these to financial information. If practitioners feel the pressure to act against their own professional judgement and ethics, the way they interpret certain events might shift away from the perspective of the evaluator(s) of the records. It is assumed that accounting education and training would equip the practitioners to withstand any pressure exerted by the management to develop a narrative of organizational performance that differs from how it really is, and that they can rely on guiding principles, the accounting standards, to develop an independent narrative of the financial reality of the firm. Accordingly, accounting as well reporting practices are considered to be important devices for discharging accountability (Connolly and Hyndman 2004).

In the realm of public sector and governmental systems, where the common public (taxpayers) is directly interfaced with the governance structure (including accounting), accountability is viewed as a responsibility to offer transparency into accounting and to acknowledge the participatory role of individual taxpayers within the system. Although governmental accounting has not always followed private sector accounting standards, it is based on the concept of accountability, primarily in deciding public policy and in its execution through the budgetary process. For example, in evaluating the accountability practices of Porto Alegre's participatory budgeting process, Célérier and Cuenca (2015) demonstrate how an accounting–accountability process involving citizen participation in the budgeting process affects the political field and the rules within which politics is conducted. Although not entirely relevant to our discussion in the role of accountability from an accounting perspective, it shows how accounting and accountability explicitly overlap in public accounting systems, while being implicit in the case of private enterprises.

WHO ARE FIRMS ACCOUNTABLE TO?

This brings up the question of whether firms exist *only* as an instrument to fulfil the economic aspirations of society or whether we can also view them as the social institutions that the critical theorists would like them to be, which ultimately relates to the accountability of firms and their enactment of accountability within the competitive world of businesses. To me, the definitions of accountability and its role in firms, both from prescriptive as well as descriptive perspectives, are narrow and unsettling as the previous section detailed. But is this question even important? Let's explore. As per the legal definition of a private enterprise or company, the accountability of firms is controlled by the relevant legal structure that sets the legal bounds of its operation and governance. This sets out the behaviour of firms in a prescriptive manner, so that any breach in its behaviour that goes beyond the perimeter of law (insider trading, inaccurate financial statements and deliberate concealment of information, to name a few) is handled through legal recourse. Still, there are other occasions where within the bounds of legality, corporate behaviour is felt to be improper at some other level. Examples include corporate outsourcing to sweatshops (examples include the carpet, garment and tobacco industries), oil spills in water bodies (the Kuwait oil fires and leaks, the Gulf of Mexico spills), lack of rehabilitation of people displaced by mega structures (large dams, land acquisition and grabbing for industrial and national projects) and industrial projects in protected areas (for instance, those with fragile ecosystems, with historical and religious importance, and where biodiversity is affected). Such examples are countless and only a few have been named.

In either case, the impacts of these events are felt not only by those who are directly connected to it, but by society at large as well. More than the transactional aspects of the disturbances, it is the *trust factor* (trust that society has placed in the firms) that takes a beating, not to mention the aftershocks that relate to the size and status of the firm. This creates the need to incorporate these failures as a part of the social learning process. Risks associated with such failures result in incremental learning that remains too fragmented to offer a prescriptive framework within which these failures can be examined. Social learning from cases like these result, at times, in new legislations and regulatory standards as well as social overheads that others are expected to pay for (such as additional cost of compliance for firms, monitoring and report costs, cost of maintenance, and others). At the same time, societal expectations in the form of an unwritten dependency on the market economy and admiration of the entrepreneurial spirit as a system slip a notch with such failures, chipping at the foundations of *trust* in businesses. While analysing these situations is one part of the story, realizing our inability to predict or fully comprehend the impact of social miscarriage is another, which incidentally relates to the absence of a suitable background against which these failures or fractured states can be explained. These issues severely damage the reputation of businesses as responsible institutions, at a time when flourishing market economies around the world are signalling the success of the entrepreneurial spirit. However, the question that remains to be answered is why this should be a problem at all if no legal breach is observed. In all fairness, is society overstepping its bounds in judging the conduct of businesses or are businesses being as responsible as they should be? It is no surprise that the available scholarship on the accountability of firms is an unfinished theory that future research must contribute to.

Current theories on corporate accountability are challenging, to say the least. While firms have been granted the status of 'persons'(artificial though), scholars and theorists have relied on the social theories of ethics and justice to define accountability, and its application has been limited to defining behavioural perimeters for employees and managers. In other words, research covering individual ethics and morality, employee behaviour, and managerial ethics and duties have been used extensively to explain 'organizational accountability', using the former as a decoy to frame the latter (the normative view). One argument in support of this strategy could be the underlying assumption of companies as groups of people ultimately coming together under one umbrella to work towards a common cause and shared goals, so individual accountability of employees (prominently of its top management) somehow converges to organizational accountability. Can aggregated individual accountability substitute organizational accountability? Here is my analysis.

Let us begin from the beginning, by analysing corporate behaviour first (if such a thing exists), followed by the ethics and accountability of firms (scholarship in one area is mostly non-existent and in the other, it is narrowly focused). This will help to situate the firms' behaviour against the context of societal expectations.

While the societal expectations from firms would be that they learn, grow and behave with increasing responsibility in successive generations of organizations, the market and economic expectations remain grounded in performance and valuation, creating a mismatch. In an era where we have witnessed a federal government (that of the US) bailing out failing enterprises under the guise of them being 'too important to be allowed to fail' in the recent past,[1] going against the fundamentals of market economy that demand letting firms follow their natural course of existence not only dilutes economic forces in favour of governmental intervention, but also represents a moral hazard. If these breaches were mere anomalies, they would still have been acceptable. However, organizational failures due to unethical breaches and resulting from moral turpitude need a referential framework against which they can be analysed. While ethical behaviour and moral obligations at the individual level (of employees and managers) can be guided by aeons of scholarship to help us behave as and become better persons, it cannot compensate for organizational ethics or the lack of them. In the absence of any prescriptive frame to shape organizational behaviour, the cost of learning would rise with every ethical breach, including the cost of restructuring institutional set-ups through the unlearning and relearning processes that are believed to be the best practices of corporate governance. This is more of a business-as-usual approach.

Let us now transition to the discussion of ethics.

The top management of most of the successful organizations today are products of business and engineering education, which breed a unique learning process where success is defined through quantitative measures that the market—or shall we say, society—prioritizes, institutionalizing them as the default for the education systems to build on, followed by the media drumming these values and governance adopting complementary policies, brushing aside critical and challenging perspectives on 'success' that the newer economic scholarship and market systems are bringing in. No wonder management schools considered organizational success and employee engagement as cogs in a factory system where the only form of ethics that exists is the one that echoes the successful practices of the top management of large multinationals. This narrow focus in defining a successful organization also confines the role of ethics for top management to a narrower theme, taking the focus away from the core concerns of organizational

---

[1] The federal takeover of Fannie Mae and Freddie Mac and promulgation of the Emergency Economic Stabilization Act of 2008 to bail out financial institutions are examples.

existence—that is, whether organizations are expected to evolve ethical behaviours and characters in due course of their existence.

Companies have already devised charters of prescriptive behaviour for their employees, contractors, independent directors and in some cases CEOs as well. Moreover, ethical failures on their part are covered by ethical and legal frameworks, but that can hardly replace the need for establishing accountability for corporations in their interactions with society and the environment. A part of this accountability, without doubt, can come from managerial and employee-related ethics, but cannot be the sole way to shape it. Moreover, learning from social psychological theories and group dynamics, we know that the chances of coercion by authority are real and irrefutable (refer to the Milgram experiment: Milgram 1963; refer to the #MeToo movement in the corporate arena in 2018) and necessitate that firms develop an ethical code, beyond the bounds of legal frameworks. As you have noticed, I have used 'ethics' to refer to accountability in this part of the discussion.

Let us now consider accountability specifically.

Accountability, as a part of ethics, relates to the behaviour of individuals in relation to others and the self, regulating how one behaves and holds oneself responsible to others and the self. Translating this to the institutional level forces us to develop a prescriptive view (or a normative view) of how an organization should behave towards others and itself and how it should hold its own conduct accountable through the actions it takes (beyond the realm of economics). This brings in the behavioural elements of 'ought to' instead of just 'should' or 'could'. Since a firm is an artificial person, the rationality of its behaviour cannot be connected to the collective or generic learning processes but could improve from having a supranational framework. Research involving organizational accountability has overlapped with business ethics, where the roles of employees, managers and top management in shaping the organizational climate to induce ethical behaviour converge to displace accountability within organizational ethics. Here, accountability gets embroiled with other ethical considerations such as responsibility, honesty, respectful behaviour and others, instead of taking a stand of its own. While it is easy to propose how each of these should operate in isolation, it is challenging to pick one over another in a given situation.

As the critical theorists are demanding organizational accountability with regard to social and environmental concerns, their line of argument (as detailed elsewhere) is that the organization should demonstrate accountability with regard to social concerns as responsible corporate citizens, which to me seems more like a search for corporate empathy for their less successful brethren, for society and for the environment. This in turn is like searching for empathy in the psyches of rich nawabs, who are so unfortunate as to have not experienced the uneven world they are a part of, including never suffering from hunger and pain. What would motivate them (the Nawabs) to share their riches with the society, as we would agree that by donating to the needy, we satisfy our altruistic needs and not necessarily uplift the lives of recipients (although we would hope so). While I leave the intents of such kind acts and related debates in the able hands of behavioural psychologists, a question that merits attention here is why firms should be interested in investing in the means and methods to improve their own accountability that has neither any mention of returns, nor any definitive means to ensure accountability. Moreover, in the absence of any norms to govern how a firm's 'character' should evolve in light of changing social expectations, why would the firms find this quest more important than other competing urgencies or evolve such norms on their own? At the same time, it would be equally unfair to hold them responsible and accountable for something that they have not been designed to evolve, especially where the societal expectations are not well articulated. Lastly, accountability as it refers to firms is yet to find a clear expression that firms can debate and agree upon. Next, we explore the normative theories from the school of ethical thought, to deliberate what accountability, in its institutional role, could mean for firms, and draw a boundary to examine business conduct of firms.

Ethical theories and business entities

As discussed earlier, the role of business ethics is oriented towards organization building and less towards letting corporations (as a social unit) reflect on their behaviour. Scholars have lamented that other than discharging accountability through accounting information and disclosures, most of which are financial or economic in nature, there is hardly any other form of accountability that has been institutionalized for firms. Critical theorists have brought in the angle of social and environmental concerns and society's expectations for firms to appreciate these concerns and behave differently from how they are doing today, but without offering any form of prescriptive framework (to be discussed later), hoping that the firms will take cognizance of the shortfall on their own and amend themselves. At the same time, most of this is related to ethics and does not deal with accountability per se, though the interchangeability of these terms is accepted in the literature, more so because the accountability of firms is yet to be defined beyond the existing, narrowly institutionalized version.

Introduction to *ethical approach* to business is theorized for firms to view its quasi-public nature as the underlying reason for them to deal with the larger issues that the societies are facing today, instead of waiting for a formal institutionalization of ethical behaviour. An ethical approach advocates that firms should view their obligation towards nature and society as a primary concern and find ways to evolve an uncompromising attitude in business to uphold this value as a central tenet of their existence, simply because *this is the right thing to do* (the normative view). Even though this view may not always lead to a win–win situation, especially when the outcome(s) of any strategy or opportunity is/are stacked against the economic ones, firms are expected to align their actions and decisions to maximize the overall well-being of everyone concerned. However, not only would it be impractical to get into the calculative aspects of enacting such a vision, but such an ideology is a difficult one to follow even from a pragmatic standpoint. Still, a notable aspect of this approach is that it leaves the choices to the firms to act on, hoping that they will go beyond taking a utilitarian view of their decisions to develop an emancipatory nature and gain maturity in due course, enabling the firms to eventually handle the praxis of different dimensions with ease. This is overly wishful thinking, perhaps, without offering firms either a language to express their intentions effectively or supporting them with proven methodological improvements. Either way, the complex nature of such choices could prove costly to firms, and confusing as well, at times.

Even though following this emancipatory approach closely corresponds to the 'deep green' approach (caring for nature in everything we do), it is contrary to the methods of economic organizations, where economic considerations generally outweigh every other option. At the same time, this kind of approach goes against the progressive adoption of environmental and social superiority that the incremental or evolutionary approach suggests. To exemplify, the inadequate methodological support of the current knowledge base in evaluating the holistic impact of doing business and accounting for concerns such as biodiversity loss, permanent loss of natural capital, reduction in inter- and intra-generational equity and loss of ecological resources leaves firms simply overwhelmed and unable to react effectively. Although scholars have relied on the theory of natural capital to connect the built environment and sustainability and to account for the contributions in natural and manmade wealth, the absence of a demarcated boundary or a causal relationship as well as ownership issues between the two introduce problems of accounting and technical complexities that call for redefining accountability. It is comparatively easier to voluntarily report the organizational response towards these concerns which the firms may have already adopted. Although a detailed examination of disclosure practices has been undertaken elsewhere, suffice it to say that indiscriminate reporting by firms does not reflect increased accountability, but rather, more of a business-as-usual approach that adds to misinformation.

Experts have cited the theory of ethical rights to uphold stakeholders' rights to be informed about how firms are handling the environmental demands of the society. Accordingly, more than the empirical aspects of the subject, we explore here the systemic ones, including examining why firms have not always had a good track record. As per the ethical rights theory, the ethical negatives generated by firms are not to be set aside as the implied outcomes of conducting business, as the propensity to highlight only the ethical positives is comparatively higher. Fundamentally, ethical negatives emerge as a response to the corresponding positive right (e.g., the right to life can also be viewed as a negative right *not* to be killed)—they cannot be considered as belonging to mutually exclusive domains and instead develop logically to counter each other. To apply this to business would mean that a firm's licence to operate is based on its perceived ability to add value to society beyond what individual contracts could have generated (per the transaction cost theory of firms), but that would also mean that firms must remain transparent and open to scrutiny by the society, beyond informing stakeholders alone, and all the more so in the case of negative impacts and the steps they take to ameliorate the situation.

Management theories and ethics

The only area where management theorists have been able to contribute to improve the accountability of firms significantly is the stakeholders' theory, which improved the accountability of managers beyond shareholders. Riding on the backbone of the agency theory, where the role of the manager has been enlarged to be accountable to entities beyond investors and others providing financial support to the firm, the stakeholders' theory directly enforces managerial accountability for wider considerations of actions, thereby leveraging the legitimacy theory to demand that the actions of management be in accordance with the responses of a wider audience. At the same time, it holds managers responsible for recognizing and discharging the accountability of management towards others, while letting them define what those spheres of concerns are or could be. This makes the role of managers an important factor in building trust and interdependence with society. At the same time, this is not a prescriptive framework but allows space for the concerns of relevant stakeholder groups to generate sufficient weightage and appear on the radar of the management. This calls for societal activism more than organizational activism. To me, it shows the management (or policymakers, in some cases) a way to ignore, thwart and depose minority voices. Accordingly, social and environmental concerns remain a theory, not due to intent, but due to viewing organizational accountability as same as managerial accountability, which unless translated to financial spheres and ranked sufficiently high would remain ignored.

Here is an example to expand on that a bit more: Israeli game theorist Ariel Rubinstein got the idea of examining how the tools of economic science affected the judgement and empathy of his undergraduate students at Tel Aviv University (Rubinstein 2006). He made each student the CEO of a struggling hypothetical company and tasked them with deciding how many employees to lay off. Some students were given an algebraic equation that expressed profits as a function of the number of employees on the payroll. Others were given a table listing the number of employees in one column and corresponding profits in the other. Simply presenting the layoff/profits data in a different format had a surprisingly strong effect on students' choices—fewer than half of the 'table' students chose to fire as many workers as was necessary to maximize profits, whereas three quarters of the 'equation' students chose the profit-maximizing level of pink slips. Why?

The 'equation' simply 'solved' the problem of profit maximization, without any need for the students to think about the employees they were firing or the consequences of their action. Rubinstein's classroom experiment serves as a lesson on the pitfalls of the scientific method: It often seems to distract us from considering the full implications of the calculations that managers use to arrive at certain decisions. The point isn't that it is necessarily immoral to fire an employee, but rather that when the students were encouraged to think of the decision to fire someone as an algebra problem, they didn't seem to think about the employees at all. How far are business managers from adopting this fallacy? Such deviations shed light on accountability from the institutional perspective.

To support the cause, I would request researchers and practitioners to refer to the ancient Indian concept of dharma, which I believe defines accountability most accurately. *Dharma* in Sanskrit is not just religion; it also translates to holding oneself accountable to and for both self and others, while being righteous in performing duties, in behaviour and in conduct. By this definition, accountability is not about conformity but about righteousness, about being able to hold oneself responsible for the outcomes, which nearly closes the gap between 'could' and 'ought to'. This relates not only to individuals, but also to institutions and institutionalized practices—for example, as defined by Kautilya[2] for kingdoms and as the king's 'raj dharma', the gold standard of kingship. Dharma is not a one-dimensional feature of behaviour but holistically and deeply rooted in the well-being of all concerned as against benefitting one person or primarily one group, which is closer to the ethical approach we are looking for and contrasts economic principles, the foundation of modern business enterprises.

I would relate the foregoing discussions to the shortcomings of the current version of accountability, which has remained confined to the obligations that the preparer of the books of account has towards others—that is, it is applied solely as to a principal–agent relationship. The concept of dharma, however, is not limited to individuals but can be applied to institutions as well: Where the impact of institutional existence is much wider, so would its responsibilities and obligations towards subjects influenced by its activities and decisions be wider correspondingly. Dharma relates to individuals (as to kings, so to CEOs) but also to kingdoms or social institutions that function as a part of society, take inputs of resources as society allows and serve the interest of society (or subjects) in a meaningful way. Accordingly, there is no reason to believe that social institutions like businesses can live without being accountable to the society. Importantly, institutional accountability should not be tied to being faithful to ownership alone, but to the well-being of all concerned, according to the ethical school of thought. I would request researchers and thinkers to contribute to this endeavour and enrich it from the organizational perspective. A new form of institutional accountability needs to transcend the confines of what is to appropriate to accounting books and become the guiding principle of how organizations are responsible to society not only in what they produce or serve, but how they produce and serve it and the manner in which they interact with resources, stakeholders, the environment and society to do so, even as they need to earn a profit to survive. The institutionalization of accountability in

---

[2] Vishnugupta, also known as Chanakya (4th century BC), was a teacher, strategist, economist and royal advisor, a scholar of Takṣaśilā University, who authored the *Arthaśāstra* in Sanskrit, a pioneering work of political economics, and laid the foundations of the Mauryan Empire.

government-owned enterprises has been seen to follow some of these tenets, maybe because they are working with public money, whereas private firms are allowed to function with muted accountability. While this has been the way so far, it threatens the fabric of a democratic society if such an important institution is left with a limited form of accountability.

Throughout this discussion, I have kept in mind Foucault's work on changes in epistemes, referring to his work on the modern episteme (from 1800 to the present) that connects to all disciplines other than grammar and language. Using epistemic analysis as the archaeology of the human sciences, we refer to the development of knowledge and its relevance to human kind (Birkin and Polesie 2011), where the episteme shows how people order their knowledge and how they construct reality and develop their worldview. It defines what and how they get to know of their world and themselves. It influences what they value and how they make meaning. From this perspective and the incompleteness that knowledge from any given period might carry if applied to analyse and build suitable understanding of situations across generations, my intent is to seek a cross-pollination of ideas, where an idea from the past can have a certain relevance in the current age that we can lean on and use to construct new knowledge, using the theoretical lens of modern times to analyse it.

Financial Accounting, Reporting and Accountability

INTRODUCTION

Continuing from the previous chapter where we explored accountability from the perspective of economic theories and its natural overlap with accounting—which incidentally is the only institutionalized form of accountability for firms—this chapter presents an in-depth inspection of the current form of financial accounting and reporting, by reviewing multiple aspects such as regulatory mechanisms, the role of standard setters and accounting practitioners, and in ensuring corporate accountability, and how all these actors have created an unequivocal control over what accounting should be (the normative view). This also leads me to examine the limitations of the mechanical view of the accounting construct, corporate reporting practices and the role of practitioners in developing narratives that generally follow the management's version of reality. At the same time, this is not a discourse on financial accounting or a technical evaluation of the accounting process. Instead, it is a critical assessment of how financial accounting and reporting practices act as signal mechanisms for the market and how they shape the market's views. This has been substantiated through empirical validation, notwithstanding the interpretive nature of corporate reporting. To develop a balanced perspective on this, normative as well as critical frameworks have been used as the theoretical lens. The analyses confined within the limited space of this chapter sits against the backdrop of how the debate on sustainability is questioning the art and practice of accounting, in response to the new demands of market and society. This also raises questions on the current movement towards the unification of global standards of accounting and whether that would address some of the critical shortcomings that have been identified in the domain of accounting.

FINANCIAL ACCOUNTING: MEASURING AND AGGREGATING RELATED AND UNRELATED THINGS

Financial accounting is the most common form of accounting and translates the financial transactions of an entity. Fundamentally, the accounting function enables firms to report organizational performance through certain aggregates, where the measurement, recording and reporting processes are designed to follow GAAPs and statutory reporting norms. The GAAP prevailing in any country is considered the ultimate guide for practitioners to interpret business activities, where the practitioners are trained in the science of financial measurement and techniques of bookkeeping that enables them to scan business activities that the firm undertakes. The accounting process converts transaction that the entity has entered into and records the interpretations in books of accounts in a manner that can be reviewed and authenticated for verifiability, accuracy and diligence by auditors, enabling them to certify that the translation and reporting process has followed the GAAP (discharging accountability).

Once the accounting information is disseminated, the user of accounting information can refer to the accounting artefacts, reconstruct the business events that the firm has undertaken over a period and explore the major financial milestones. The acid test that this process passes through is repetitive deconstruction of accounting information by different users at different points of time to reproduce the financial behaviour of firms with reasonable accuracy, by referring to the GAAP applicable to the period and the place of business. So while the accounting practitioners use the GAAP as a general guideline to translate business events into a specific measure of accounting, thereby producing one version of reality, its subsequent deconstruction by the user of information is expected to generate the same version of reality or a closest approximation. To remove interpretive bias, accounting standard-setters in the USA (the Financial Accounting Standards Board, or FASB) have historically practised rule-based accounting, which directs (rather than just guiding) the accounting practitioners to interpret events in a *specific* way, such that the interpretations should have a minimal chance of misrepresentation. In other countries, standard-setters have adopted standards that are based on the international accounting standards promoted by the International Accounting Standards Committee (IASC) and improvised upon these to develop country-specific GAAPs that suit local reporting needs. At least, this is how it was till the end of the 20th century. Since then, the IASC has been working to promote a universal GAAP—the International Financial Reporting Standards (IFRS), which are principle-driven, maintained by the IASC and towards which country-specific GAAPs are converging.

In the US, the Securities and Exchange Commission (SEC) issued release nos. 33-9109 and 34-61578, the Commission Statement towards Convergence and Global Accounting Standards on 24 February 2010, paving the way for the convergence of the Accounting Standards Codification (ASC) and the IFRS. However, the process of convergence is still underway, in the USA as well as in other countries (other than Europe and countries where the IFRS has already been adopted in the last decade). At the same time, we can see material differences between the ASC versus corresponding IFRS standards. Take, for example, IFRS 15—the accounting standard governing accounting for revenue from contracts with customers—and the FASB's attempt to converge on it, in the form of ASC 606, originally published in May 2014. The latter standard covers accounting for revenue from contracts with customers except for the ones covered by lease contracts, revenue from financial instruments, insurance contracts, guarantees and non-monetary exchanges, amongst exceptions. However, in 2016, the FASB and the IFRS issued separate amendments to clarify their respective guidances. Details of these can be found from practitioner-oriented disseminations.

The principle-based accounting standards that the IFRS promises are expected to improve transparency, accountability and efficiency. The IFRS relates to transparency in terms of international comparability and quality of through standardization of artefacts, whereas accountability relates to reducing the interpretive gap between management and market. The efficiency that the IFRS is concerned with is about helping investors identify opportunities and risks (IFRS 2018). Having said that, principle-based accounting standards are believed to place a higher emphasis on practitioners' judgement in interpreting business events and translating them to the equivalent accounting information. This addresses concerns about the firmness of the construct that straitjackets the outcomes of the accounting process and allow the practitioners to search for a better way to interpret the transactions, if the real events have not followed the trajectory that the standards expected it to follow.

Having said that, this discourse is not about deliberating on the merits of one GAAP versus another, other than to allude to the fact that the systemic inheritance of accounting in translating the performance of firms remains confined to the economic realm. This includes problems of aggregation that are based on certain basic principles of the measurement system but are thrown together in accounting statements. For example, cash is measured with a high level of accuracy, but measurement of goodwill impairment could be inaccurate, given its dependency on the impairment principles, whereas these being current assets, are aggregated together in the balance sheet. It is to be noted that accounting aggregations depend on how individual components are measured, weighed and included in or excluded from an accounting group, and not on their fitment within the said group due to homogeneity and nature of similarity. At the same time, when companies combine within a group to report on their consolidated performance, such aggregation—for example, the accumulation of goodwill as well as the removal of inter-company profits that are measurement-dependent—could result in misleading information. These situations cast doubt on the accuracy of the accounting information or the mental map that the users of information create, although they have relatively less impact on reliability as the same set of information can still be produced by following an identical set of processes.

Before exploring the systemic issues, a couple of valid questions are how distorting such impacts are and whether having more rules would help. Isn't it possible that incomprehensibility and the risks of accommodating more rules would result in an increased number of notes/footnotes in the accounting reports? For example, by allowing fair value reporting of assets and liabilities, a trend that is on the rise, we have allowed greater buoyancy within assets, liabilities and earnings that, although wash out over a period, leave intractable marks on the market valuation of stocks and the risk ratings in the interim period. Similarly, in this era of technology, unbooked intangibles (such as knowledge capital and goodwill) are overtaking the value of tangible assets (such as buildings and furniture). This complicates matters further where the value of the intangibles collectively outweighs the tangible assets, not to mention causing a volatility in the valuation of intangible assets that is concerning, as compared to traditional accounting that used to be more concerned about the tangible assets. Should we consider changing our GAAPs or devise mechanisms that create objective information for the market?

Value realization of firms

Another point relevant to this discussion is the absurdity of value realization that financial statements under the GAAP are expected to provide. There have been no good models to describe value creation by firms in the marketplace and to later realize the value that has been created by a firm. Value creation in the era of industrial development was related to the earnings or profits generated by the firm (the profitability or earnings view)—the ratio of market value to book value used to be 2:1, which has now moved to X:1, where X can be anywhere between 6 and 100. So what is measured in the balance sheet does not adequately reflect how market values are created or whether the calculated market value accurately reflects the worth of an organization at a given point of time. Accordingly, a financial statement based on the GAAP does not provide sufficient information to investors to decipher which firms (stocks) are the true value creators versus those that show short-term buoyancy. In a short time frame, this might not be an issue, but over a longer span, investment opportunities and the flow of capital might change direction due to a shift in the flow of investment towards the wrong candidates, while the economy misses out on the best allocation of resources.

Another area of concern in this regard is the inability of the market to shape sunrise industries or channelize excess valuation at a gross level to the new industries, which might help the economy to move excess or surplus funds towards developing spare capacity for tomorrow. This is not about creating a blue ocean for new entrants. Rather, this relates to the systematic planning towards capacity augmentation that needs to take place to help industries relate to the economic vision of the nation–states and raise the funds necessary to rebuild capacity for the competitive landscape of tomorrow. In exploring company dissolution (winding-up), I have touched upon aspects of business and opportunity losses due to the early removal of potential organizations in the sunrise industry where standardized accounting processes and practices hardly allow new firms to demonstrate their capabilities beyond numbers, which are not reflective of the dynamic needs of the marketplace of today.

Better handling of these concerns has been seen in the defence sector, particularly in the USA, where technological innovations have been incubated over long time horizons and harnessed later for strategic and economic gains. However, the defence sector in the US, as also elsewhere, does not need to publish financial statements for market consumption or worry about the valuation of incubated innovations. The point here is that, in the absence of scientific information explaining why and how the reported information connects with the blueprint for the future in the short term as well as in the long run, a surge in the markets can run on sentiment and rumours, resulting in volatility, which eventually would result in a higher cost of capital, as investors would demand to be compensated for the volatility. A systemic push towards increasing the cost of capital is challenging unless we can find out a way that is relevant and able to improve the objectivity of information being disseminated through the GAAP-based standards and/or newer mechanisms that are more objective and accessible. Merely changing the GAAP or adopting a new set of standards might not be the answer unless there is a process to debate and promote new ideas before promulgating new standards. Otherwise, redundancies from one system are bound to seep into another, allowing them to acquire deeper roots in time.

This brings us to the need to explore the relevance of financial statements and their dissemination, including their ability to approximate the value of a firm. Since the landmark publication of Ball and Brown (1968) that provided empirical evidence connecting reported earnings with movement in the share prices of firms, the argument that the market valuation of organizations depends on the performance of the firms (read: their profitability) has undergone revision.

First, we consider the income or profitability view. Ball and Brown (1968) correlated the fluctuation in the share prices of companies to their reported earnings, thereby questioning the conventional wisdom regarding the usefulness of information on earnings based on historical costs beyond a statistic. The reference point of valuation in this case was based on the reported earnings of the firm. Net earnings and earnings before interest, taxes, depreciation and amortization (EBITDA) are important metrics because investors change their portfolios based on net earnings and its derivatives more than anything else in the balance sheet (Verrecchia 2013). I believe that managers have a functional fixation on the earnings metrics—not because they believe that other managers and investors prefer these because of their monumental importance or because they are inherently great, but because they believe that for the market, this index is of vital importance. Accordingly, in defining income, managers are averse to including items that are not performance-enhancing. Excluding items that impact income statement leads companies to advertise their marketworthiness, while standard-setters such as the FASB and the IASC can control what goes into defining net earnings. The same is the case with EBITDA, which (as a part of the income statement) earlier used to help companies report on their income, even with discretionary elements attached to derive the net income.

This brings in the balance sheet view. Since Ball and Brown, the emphasis of reporting has shifted from income statements to balance sheets and, in accounting, from historical valuation to fair value accounting. Saito and Fukui (2016) posit that over the years, the concept of income in financial accounting has eroded in favour of the balance sheet approach that the standards-setters (the IASC and the FASB) have followed more vigorously in the second half of the last century. They believe the IASC has prevailed with the asset–liability approach as information on the fair value of assets and liabilities contribute to corporate valuation, and future cash flows became a function of the changes in the levels of assets and liabilities. This change, the authors argue, can more logically be connected to derive revenue and expenses from the recognition and measurement of assets and liabilities. This view collapses the future earning potential of the firm to its equivalent present value and in favour of other measures. Moreover, the balance sheet approach ignores the role of intangibles such as goodwill and changes in the fair value of assets and liabilities that cannot directly be connected to the cash flows but can eventually contribute to the valuation by influencing the perceptions of the market. This challenges the view where the market worth of firms is evaluated in accordance to the market value that they would fetch, should they be forced to dissolve at a given point of time.

Even when we persist with the argument that such an approach helps investors with stock performance in acute cases such as bankruptcy, this is not in fact the case, as we will discover and explore in the last section of this chapter. In research that comprised panel data across 40 years (1962 to 2002), Beaver, McNichols and Rhie (2005) examined a select set of US companies to find that over the years, data submitted by the companies to regulatory authorities had become less significant in predicting bankruptcy. This needs us to examine the hard truth where it matters most: How can economic methods and the accounting focus help manage risks and bankruptcy, signal a downtrend and help firms turn around. We will discuss this next.

TRIANGULATING ECONOMIC METHODS AND THE BUSINESS RELEVANCE OF ACCOUNTING FOCUS: THE CASE OF BUSINESS FAILURES

In this segment, economic method and accounting focus are brought together to reflect on the current stock of knowledge where it matters most: the endgame, or business failure. Researchers in this area have generalized failures as a part of the life cycle theory of a firm, extrapolating the existence of firms and the challenges that a firm might face in different lifecycle stages. This places organizational evolution against the canvas of a life cycle that is easier to comprehend. However, firms being artificial entities and a product of economic, competitive and technological frameworks, their survival can be better explained using PEST (political, economic, social and technological), SWOT (strengths, weaknesses, opportunities and threats) or PESTEL (political, economic, social, technological, environmental and legal) analysis, these being common frameworks to explore how a firm navigates through the competitive landscape. However, it remains difficult to generalize about business failures, as we shall see from a detailed examination.

Before continuing, presenting the rationale for selecting business failure as a cornerstone for this discussion is also warranted. From the pool of dissolution cases, what is *not* being covered here includes:

a. Dissolutions of companies due to unethical practices and voluntary financial irregularities,
b. Business discontinuation due to a merger or acquisition,
c. Voluntary dissolution that is ordered by the law of the land or is due to changes in the regulatory systems, and
d. Shock elements and entrepreneurial learning (from behavioural and cognitive perspectives) due to the failures.

The coverage includes business failures that are not only related to bankruptcies, but also insolvencies within the competitive landscape. At the same time, our intent is not to theorize on the why and how of the dissolutions, but to derive generalizations on the existing mechanisms of accounting and reporting systems and their bid to provide information to the stakeholders, if available, and help firms enact their accountability to society. From the accountability perspective, critical business issues should form a part of the public narrative and contribute to discharging their obligatory function towards stakeholders that the ethical rights theory demands.

From the accounting perspective, one of the underlying assumptions is treating a firm as a going concern, whereas the life-cycle model of firms depicts business ventures as passing through the life-cycle stages of evolution, maturity and dissolution, signalling a fundamental disconnect between the economic and life-cycle views. The dissolution of firms according to the life-cycle theory of firms is based on the theory of natural existence, suggesting that firms move from one stage of life to another before becoming terminal cases, which for firms is less related to ageing and more due to adaptability and a changing competitive landscape. Although we can safely assume that the experiential learning of the firms is helpful in successfully navigating economic turbulence, in reality the dissolution of business is always a possibility and a function of operational complexities, beyond the life-cycle rationale (setting aside unethical practices like misappropriation of funds, accounts falsification and other malpractices, which anyway are a recipe for disaster irrespective of whether they get reported immediately or not). However, the disconnect between the two views is not what is being investigated here, but rather how the accounting view of a company's affairs gets communicated to the market. There should be some means to communicate any impending concerns that a firm might be facing. Since financial accounting and reporting operate as a signal mechanism for the investors and lenders in the market, there should be a way, one would argue, to signal distress when a firm finds it hard to make its way, instead of reporting only 'business as usual'. This would induce stakeholders to extend all help, one would hope, to revive the firms and reach out to wider agencies, instead of waiting till the terminal stage (filing for bankruptcy).

Lane and Schary (1991) have pointed to structural changes as the primary reason for business failure, and Dun & Bradstreet is one of the organizations that routinely collect this data. Moreover, the authors say, younger companies are more susceptible to structural changes in economies, specifically recession, where their failure is viewed as being essential to redistribute scarce resources in an efficient way—but this is the economic view. In an analogy to human existence, children are protected, sheltered and groomed till they are of age and can fight for resources on their own. Instead if societal aspirations are not protective of start-ups, this simply translates to a survival barrier, maybe because the collective experience of society reflects this test as a mandatory one for long-term stabilization. Another way could be letting the market forces decide if they are open to accepting the new ideas, themes or competition. There cannot be, to my mind, a third theory to explain why start-ups cannot survive in a mature industry, other than to believe that society is not ready to bear that responsibility and hence interferes with market forces.

In all fairness, firms operate in an economic environment, and draw its life support system from the society that it serves and from the environment that provides for all the natural ingredients firms need to create, generate and prolong its existence and market value. However, there is no underlying theory to connect firms with the value they generate other than the economic ones, which incidentally limits our understanding to reasonably explain challenges firms might be facing in case of a sudden deterioration in its value creation capabilities. One reason for this limited understanding could be information abstraction, that is limited to financial information, disconnected from the thick description of business and the social reality within which the firm is operating. From the scholarship on this approach, the dissolution of firms and related research has generated interest in the last century, pursued mostly in the US, due to the well-structured market and regulatory system that have defined clear entry and exit paths for firms to follow. However, the research on this subject has also covered other markets as well.

In other reviews on the subject, the dissolution of firms has been considered from the financial distress perspective, which relates to the inability of the firms to meet their financial commitments (credits, loans, etc.), and accordingly distress is understood as the firms' inability to generate enough cash to meet their requirements (Beaver 1966). Failing to do so directly relates to their inability to service debts and fund operations, which signals operational distress. Some salient points of these other perspectives and frameworks follow:

- Cash flow approach: Models developed from this theory used financial ratios such as cash flow to total assets, net income to total assets, total debt to total assets, working capital to total assets, and current ratio to evaluate if any one of them signals the affairs of the firm becoming out of the ordinary. Accordingly, research in this area used methodologies such as univariate statistical analysis to assess the predictive ability of the approach. This approach was improved by Altman (1968) to use multiple discriminant analysis (MDA). Over the years, research has advanced by examining different cash-flow measures and by bringing in fund flow-based models (Kahya and Theodossiou 1999) to predict bankruptcies.

- Fund flow approach is based on the tenets of better earnings, large investments in the future (capital expenditure), smaller expenditure of fixed funds (long-term debt) and larger liquid assets to reduce the probability of a firm's failure. Most of the research from other international markets followed MDA, with improvisations in the measures selected, and embraced the predictive power of the model. At the same time, most of the research in this area agreed that there was a wide discrepancy in the variables used to predict success versus failure and developed models based on non-financial factors as well, which the researchers felt worked more appropriately in the case of small businesses.

However, all these models are based on the survival outlook of the firm, as the firms would have to compete for resources that are much more challenging to obtain in the formative phase of their life cycle, as compared to in their growth and maturity stages. For example, Lussier's (1995) model included 15 variables to predict the success of businesses in their early stages, some of which also covered inputs from entrepreneur's background. However, most of these models concentrated on the financial viability of the enterprise and worked with financial ratios to improve the predictive ability of the models. From the methodology perspective, this improved the outlook by adding risk-index models and condition-probability models such as the logit, probit and liner probability models.

Lee and Han (2012) noted that financial ratio-based models are stationary, as they inherently represent the ratio of two variables at a given point of time. Advanced arguments in favour of developing time series-based models of some of the important ratios, with the premise that the movement of some of these ratios over a time period can result in improved predictability. They used a quartet of ratios—that is, return on assets (net income to total assets), liability ratio (total liability to total assets), working capital ratio (working capital to total assets) and current ratio (current assets to current liabilities)—over a period of time, challenging the earlier wisdom of using these ratios that introduce stationary property, for example, in Altman's model of 1968. Lee and Han checked the trend of the time series of sample firms so as derive a pattern by isolating the time series with the unit root (the random walk model) from the industry to which the sample firms belonged; the latter cannot be discerned individually.

Their final findings remained inconclusive and could not relate the firms' performance to that of the industry to which they belonged. However, the experiment added to the theoretical backbone that organizational failure is not an isolated event at a single point of time; rather, it happens over a period of time and the organization may or may not share the same fate as the entire industry.

Conducting a co-citation analysis, an analytical method to identify influential and highly cited authors and publications in an area, Kücher, Feldbauer-Durstmüller and Duller (2015) analysed 309 articles in two different time periods (1935–1999 and 2000 onwards) to collect the prominent views on the failures of firms. This seemingly obscure conference article has done a commendable job of collating the articles, books and scholarly literature on the dissolution of firms and has gravitated towards three clusters, which included prediction-based models (discussed previously), finance- and law-related risks (bankruptcy laws, regulatory compliance and financial distress), and organizational failure (an approach to relate to business failure inside-out). It is the third cluster of 99 publications that is of particular interest here, within which only a handful of publications have analysed insolvency from perspectives that are different from the mainstream economic or finance-related theories. From this cluster, the seminal concepts and innovative ideas reflecting how an organization might fail are abstracted here. For example:

- *Liability of newness* (Stinchcombe 1965) is a concept mainly about the difficulties of bringing something new to the market as opposed the existing one, and is often experienced in the cases of start-ups, new product lines and ideas, and so on,
- The evolutionary dynamics of industries are expressed by the theory of similarities (Boone and van Witteloostuijn 1995; Peeters et al. 2014), which relates to the *inertia of similarities* (Hannan and Freeman 1984), where any differentiation would need energy and a push to get it moving,
- *Liability of obsolescence* relates to the inability of firms to adapt to or change structural arrangements (Hannan and Freeman 1984; Aldrich and Auster 1986),
- *Liabilities of smallness* is the theory that new organizations face more problems than large organizations due to fractured legitimacy and reduced steadiness (Bates and Nucci 1989; Brüderl, Preisendörfer and Ziegler 1992).
- The *resource-based theory* explain how firms find it difficult to adjust to the competitive demands of the market, creating a gap between market expectations and what they can achieve with the dwindling resources at their disposal (Thornhill and Amit 2003).
- *Resource-dependence theory* counters the perception of the dominance of a rational actor suggested by the economist' views in favour of a behavioural approach to describe decision-making processes, intra-organizational coalitions, and routines that exist within organizations and to explain why failing firms cannot wield enough power to move these internal structures around to adapt to changing market conditions (Cyert and March 1963; Pfeffer and Salancik 1978).
- *Organizational learning theory*, considers failures from the perspective of what can be learnt in individual areas, allowing firms to improve their collective knowledge. Thus 'intelligent failures' were proposed by McGrath (1999) as those that could be allowed by firms as the cost of learning if it would improve the chances of future successes.

While organizational failures can be explained using an outside-in approach that hinges on economic and social perceptions of how a micro-economic entity competes and survives in certain market conditions, it would need quantitative financial and economic parameters to derive a general theory, to spot a potential candidate for failure and to predict the chance of potential failure. Using such statistical models relies on the basic assumptions that a firm would follow the mechanistic principles on which the model is based, downplaying the fact that firms exist within certain artificial (regulatory) and natural (market and competitiveness) constraints and are not bound to follow the extracted principles of cessation. A generic model in that sense would even be more challenging to build, not because of any complexity of methods, but because our aim to derive a linear model of financial distress is situated within the paradigm of financial or economic views, without realizing that:

a. Even a financial or economic view of firms is an abstract, a mental construct that is dependent yet detached from the primary activities that the firm is carrying out.
b. Mapping this to the ratios is a second level of abstraction that generates some number at a point of time, while removing details attached to explain 'why' of it.
c. Indexing these ratios further neutralizes the associated environmental variables and removes the 'how' of it.

This leaves us with certain data that can be aggregated, sorted and worked on from the macro-economic perspective. They can signal some possibilities, if presented with care—for example, within a given period, the folding up of a number of firms (as compared to the number of new entrants), coupled with the age of the failing firms (more than X number of years), can point towards concerns related to technological adaptations, provided everything else in the legal and regulatory framework is the same. A case in point might be the US retail industry, where the brick-and-mortar segment has been witnessing a continuous decline in the last 10 years as e-tailors are gaining the competitive landscape. In comparison, an inside-out approach enables us to review the internal working of the firm, through the thickness of how things unfold, connect and disconnect, to arrive at the present situation. It details the interplay of the different elements that an organization is made up of and how things break down—something not generally explained in the existing theories, though it should have been, for others to pick up and learn from. For this to happen, we need to use a case study format to help uncover the details that Eisenhardt (1989) proposed in his seminal work of building theories from case study research, a strategy I have followed to explore the tenets of environmental accounting in Section III. Not to be able to easily untangle the complexities and interactions of different elements that shape organizational journey and decisions, especially when firms turn downhill, translates to that a framework is missing to analyse the phenomenon and associated complexities, and would have to be looked at using investigative approach, and not through some derived pointed information. In other words, we would pause and try to reconstruct the things that have taken place, to de-construct, analyse and re-construct why the trajectory of a firm did not follow the 'normal' pattern.

In summary:

- Organizational existence need not always follow the natural trajectory of life and life-cycle model, which is merely a construct convenient for human understanding. A firm can survive multiples human life terms, like so many other human institutions.
- Breakdown due to moral turpitude reflects human character and its shortcomings, which despite all theories, will always remain a work in progress.
- The failure of economic entities due to structural adjustments, regulations and legal compulsions are outcomes of strategies reflecting social and/or political preferences (in a normative way).
- In all other cases, the failure represents gaps in the existing theories. For example, if start-ups must fight for resources in the marketplace to continue, it is due to the societal preference to test their competitive mettle against mature organizations within the same industry space.
- Within these bounds, the role of the accounting and reporting system to create an appropriate valuation of the firms was found

to have no real merit and ability to predict whether the firm would cease to function at some point of time.

Lastly, there is no doubt that with operational performance and the complexities of economies, the market and social conduct embroiled in political conditions, there is little motivation for firms to improve their methods, means and thinking, to capture the imagination of consumers or to produce better products and services, and to give up the relentless pursuit of 'success' that undermines innovations. At the same time, explaining the market success of new companies who relied on innovative practices, played to the imaginations of the target consumers and obtained greater success led to a growing upheaval in the traditional theories.

ROLE OF EMPIRICAL RESEARCH IN ACCOUNTING STUDIES: ACCOUNTICS

This leads us to examine the role of the positivist theories of accounting, that Ball and Brown (1968) and the various dissolution theories of firms relates to. Empirical validation of any phenomenon means replicability of the experiment and reproducibility of the results, as well as to improve the predictive ability of results. As compared to lab-based work on natural materials, setting up the right environment and the control variables in changing social settings is a tough ask. This brings up the final topic of this chapter, which covers the importance of the contextual background and its relevance to decision-making and associated interpretive elements.

Mainstream empirical research in accounting is also known as accountics—accounting and economics together, although not everyone is amenable to bringing these two unlikely subjects together (Klein 2007). Accountics makes the roles of empiricism, statistical modelling and quantitative methods important, which in turn have nothing to do with accounting principles, standards and practices. In 1887, professor of accounting Charles Sprague of Columbia University coined the word 'accountics' and defined this as the mathematical science of accounting values. However, the modern version of the term has nothing to do with the old definition of accountics; in modern literature, it is more about accounting, politics and economics together, that is, a discipline that describes how accounting operationalizes the political economy (Suzuki 2003). Unlike economics, leveraging mathematical formulations and statistical techniques in accounting could be challenging unless the realm of investigation matches with the problem of finding or working with certain patterns.

Moreover, in order to ground the findings and their relevance, beyond testing the hypotheses by examining statistical correlations of predictor, response variables and the p-value, social scientists are expected to develop a triangulation to explain certain truths (epistemology) that are a product of social reality, unlike in scientific experiments where it is a part of natural existence to frame reality (ontology). For example, Ball and Brown's empirical experiment (1968) correlated market performance with the operational performance of firms in a market setting and an economic environment that were relatively stable and less volatile back then, compared to the turbulent market and economic conditions of today. Consequently, the message that the market rewards firms in accordance with their performance also leads to intellectual, managerial and entrepreneurial short-sightedness, where achieving a high market valuation forces companies to push for better financial performances on a quarter-on-quarter basis and has become the single most pursued avenue.

A recipient of Nobel Prize in Economics Prof Erik Lundberg (1969), while announcing the names of the winners Ragnar Frisch of Norway and Jan Tinbergen of Holland in his award ceremony speech, stated:

> In the past forty years, economic science has developed increasingly in the direction of a mathematical specification and statistical quantification of economic contexts.... [T]he attempts of economists to construct mathematical models relating to strategic economic relations, and then to specify these quantitatively with the help of statistical analysis of time series, have, in fact, proved successful. It is precisely this line of economic

research, mathematical economics and econometrics, that has characterised the development of this discipline in recent decades....

While the field of econometrics or the application of scientific principles to the problems of economics has helped economists to find hidden patterns in vast amounts of data and helped economists formulate problems as well, the challenging part is when the exercise of finding patterns gets disconnected from the underlying social realities or becomes independent of the social background, allowing economic choices to become oblivious to the situational reality. At the same time, while this obliviousness improves simplicity, it could also be misleading if the social realities have shifted over time. The case is precarious particularly where economic progress or resource allocation to growth sectors is delineated from the social aspirations, and certain agendas of growth are pursued by over-simplifying complexities. While the abstraction process hides associated complexities, it also risks implementation of solutions or policy choices that are solely based on trends. There is no doubt that correlation leads to collective amnesia towards underlying complexities, confining our understanding within the boundary of interpretive ease; but such is the nature of deductive analogies, as against putting different paradigms together (the inductive approach) to develop a holistic view of the problem. Also, this relates to how research in accounting sciences and related areas have debated to tilt the orientation of the ongoing and future research towards positivism:

- Parker, Guthrie, and Linacre (2011), in an editorial in the *Accounting, Auditing & Accountability Journal (AAAJ)*, referred to the earlier works of Hopwood (2007; 2008; 2009) and Baldvinsdottir, Mitchell and Nørreklit (2010) to relate how accounting research had been detached from its practice, and as a result, how there had grown a pronounced gap between the important pillars in the discipline of accounting. However, the editorial shifted its focus from the disconnect at the ground level to reflect on the disconnect between accounting education and practice, what is being captured in business school curricula and the real-world practice of accounting, transforming the disconnect between theory and practice to one between academia and practice, which of course is another area of concern but only one side of the story. This is always a path of choice and easier to pursue for accounting academia, but hardly an effective one in bringing large-scale social or practice-oriented change, for which more engaging approach is warranted.

- The other side of the story is the overshadowing empiricism in accounting research, as Dyckman (2016) observes, also echoed by others. He lamented the domination of evidence collection based on large samples and the use of linear models (regression-based) by accounting researchers to derive scientific truth, believing it would have only a small chance to stand the test of social reality and time. Important elements missing in these studies are views and interpretations that the researchers need to gain from insights, and the leveraging of other works that investigated the phenomenon under study and which can be used to corroborate the findings.

- Accounting scholars hardly use triangulation in the course of archival investigation, experiments, surveys and interviews to investigate the hypotheses that they are examining. If done effectively, this could add confidence in the methods and results, especially if that leads investigators to arrive at similar conclusions when using different methods, and could help the researchers to develop a thick description of what is happening in the real world, how that has been or should be investigated, and to establish findings closer to reality (the epistemological view) to better explain or (re)construct the nature of social reality (the ontological view).

- Another concern is the separation of academic research from the practitioners' world, which has been a subject of intense debate in accounting circles. Accounting being a practice-dominated area like medicine and law, it evolves around the practitioners' world and also needs research to understand how things play out in the real world so as to advance theoretical contributions. This is challenging from the theory-building perspective that academicians are often interested in, because theory cannot always be isolated from the ground reality of practice, as in the case of law and medicine.

- At the same time, academic research in accounting is becoming less relevant to advance the profession in ways that scholars have warned against—professional advancements have to be grounded in practical aspects but cannot miss the solid theoretical foundations that the subject is based on. No wonder we are witnessing plethora of new standards and reporting advances promising to help firms handle sustainability-related challenges, but less of the theoretical advances to ground these in accounting theories.
- Bipolarity of practice and academic advancements in accounting should have ideally converged, with theory and practice together promoting a well-functioning system of thought instead of remaining confined to and promoting the interests of their respective areas. Although this is not an area of investigation within the text, still it throws open the question how new accounting innovations could be taken forward and whether convergence will remain a mirage.

Outcomes from empirical investigation within social sciences is like a still photograph, an image or a single time slice of the event out of the possible 360-degree X 360-degree views of the same event at a given point; not to mention one out of the infinite possibilities on the time scale. Such an image removes the complexity and variety attached to events and gets rid of the finer details, allowing the captured version of the event to be the 'true' one at a given point of time. Mathematically, we try to connect causes with effects in a peculiar way:

$$E(y) \approx ((\odot f(x_n), f(y_m), f(z_p)....)) \quad (3.1)$$

where ☺ is the mathematical operator connecting different inter- and intra-functions, and $((...))$ encapsulates the abstraction of 'y' as E(y).

What we are missing is the relevance of this abstraction remaining confined within a paradigm and its true nature, if viewed through other paradigms. If existence can be represented as a point where multiple paradigms converge at a given moment, a single finite solution ideally should bring in multiple paradigms and related probabilities.

Accordingly, the previously abstracted view of E(y) improves to become a special case of a generalized problem $E(y)_g$:

$$E(y)_g \approx ((\odot f(x_n), f(y_m), f(z_p)....)), ((\odot g(x_n), g(y_m), g(z_p)....)),$$
$$((\odot h(x_n), h(y_m), h(z_p)....)), ... \qquad (3.2)$$

where the variables and their nature are abstracted, correlated and worked out in other paradigms as well. At the same time, if the mathematical operator ☺ connecting inter- and intra-functions in different dimensions remains the same, probably we can translate that to view their relationships in different spaces or paradigms and retain our focus on the event, instead of evaluating a function in every paradigm or the nature of reality. As any truth about human society and its practices will always be transient, ever ready to acquire new shapes or forms as societies and practices evolve in time, within this short-lived timeframe, the importance of empirical research is to augment the human understanding that is based on scientific evidence and grounded within practice, nullifying rhetoric.

In the final analysis, while the market is free to decide the worth of an organization using information that is based on its past performance, this is *only* a notional value, just an assumption. Volatility of market conditions and economic realities have taught us that the economy does not operate in isolation in the era of globalization but remains connected to the global markets and economies. Our theoretical constructs are still in their infancy to be a good predictor of a situation and they need to evolve. Failure to predict the market depression in 2008 is one prime examples of the failure of the scholarship on aggregation, whereas the inability to predict, theorise or corroborate the failure of well-established companies like Nokia and Research in Motion and now the decline of GE has raised questions on the basic theories of business as we know it today. More relevant is the road less travelled beyond the turbulent markets and distressed economies that test the strengths of business models and underlying theories, where learning from the successful firms can add to our understanding of how they managed risks and why they did it the way they did.

This raises the question of whether we believe that it is natural for a firm to fail and whether stakeholders should be able to identify whether the firm is passing through a downward spiral. Should there be information made available in the financial accounting and reporting system to help the stakeholders make informed decision and to reflect the true cost of financial resources? More importantly, if this is not how financial reporting is done, should the firms voluntarily provide information to let stakeholders make better decisions? This would lead to debate, as it should be, to ensure dialogue to better communicate, which in all fairness would help the investors assess if the organizational proposal to hold the investment is costlier than what is the current asking rate of comparable firms in the market. On the regulatory side, this would lead to questions on whether it seems inconsequential for the firms to report their state of being or whether they can be a part of the sanity check that the assurance services need to fulfil. Mures-Quintana and García-Gallego (2012) used principal component analysis on the qualitative details of annual accounts to improve the predictive models that use the ratio. Thus the need to evaluate financial as well as non-financial information is seen as mandatory to better predict the firm's future and it makes better sense, as it helps the recipient of information to fill the void in the narratives that financial information leaves. However, none of that has reached a prescriptive level yet, for firms and society to agree, evident in our inability to predict the exit trends—qualitatively or quantitatively!

Cost and Managerial Accounting

Supporting Management beyond Numbers

INTRODUCTION

Cost and managerial accounting comprises of the methods, tools and techniques that practitioners in the field use to help management with decision-making and to provide information that is grounded within the organizational experience. However, this exquisite branch of accounting goes beyond the fiduciary duty of accountants to help businesses and their decision-makers with information on organizational performance and brings together tools and techniques that can be leveraged to bring the scientific temperament to the decision-making process. This area covers three distinct elements—costing, cost accounting and managerial accounting. Costing covers the methods and techniques that cost accountants leverage to determine the costs of products, processes and services, whereas cost accounting records the transactional exchanges in the cost ledgers that are internal to the organization and determines the product costs, costs of goods sold and the inventory valuation. Compared to that, managerial accounting advances the use of past performance of organization to make decisions. This includes leveraging advanced decision-making techniques from the multi-criteria decision-making (MCDM) arsenal. By the end of the last century, cost accounting had been integrated with financial accounting to form what is now known as the integrated form of accounting, and advanced economies released regulatory controls on businesses to do away with mandatory costing records, so long as they could fulfil audit requirements within the regulatory norms, although transparency in terms of value creation and transfer became subservient to the economics of doing business. In this treatise, these advances are discussed to reflect on how internally accounted data generate more insights into the value creation process as well as cost measurement and control, while supporting a scientific temperament on the decision-making process.

COST ACCOUNTING AND ALTERNATIVE PARADIGMS OF MANAGEMENT INFORMATION

Costing and cost accounting have been major innovations in the field of accounting, since it was questioned whether the accounting framework established through financial accounting (or accounting in general) could satisfy the management's decision-making needs. This included a gamut of areas such as the costs of production and distribution, cost break-up and cost control, the relationship of costs with business activities, connecting cost formulation with the organizational value chain, and so forth. Needless to say, accounting in general and financial accounting specifically are not designed to serve the information needs of the management beyond providing trends of expenditure and income, and certain other financial information. What they lacked in general was the ability to move past performance data to decipher how goods and services generate, absorb and create value, and to scientifically connect manufacturing, distribution, services and other activities to the cost and volume of goods and services, enabling the management to improve cost control. Improvements in the costing techniques brought transparency and a causal approach for firms in connecting outputs with the inputs, processes and activities— for example causally connect profit with cost and volume levels (CVP analysis).

If we refer to any standard text on cost accounting, it would define costing as the technique of collection, classification and determination of the costs of goods and services along with the organizational value creation process, while cost accounting would be defined as the ledgerization process of capturing costed transactions belonging to a business operation. In contrast to financial accounting (which covers organizational interactions with external entities while exchanging goods and services carrying money or money's worth), cost accounting records the transactions that are internal to the firm, involving the transfer of 'value' from one internal entity to another (they could be a cost centre, product or cost object). Cost accounting emerged and was legitimized as an independent accounting technique to generate information on the accumulation of costs and to lay down norms for the valuation of products and services along the value chain by using methods in accordance with the nature of the industry in which it is being applied. Cost accounting analyses, measures and derives the costs of activities, materials and processes by referring to the historical cost of materials (purchase cost), resources (labour, machine or outsourcing costs) and overheads (indirect expenses to operate a unit of production or service) while using process maps (routing) and bills of materials (recipes) to build the relevant cost profiles (along the value chain as well as along timelines).

A formulation of costs (in the classical sense) considers direct or prime costs as the sum of direct materials, resources, and expenses that are directly identifiable and attributed to the ingredients (the bill of materials and resources) forming the products and services. At the same time, the cost of different materials and services used to support primary operations and transformation processes forms pools of indirect costs or overheads that are levied on to the products or services, based on a predetermined formula or the relationship of the activities, processes and/or resources to the cost drivers. Knowing the unit and total costs of products and services help the management with cost planning, cost and capacity utilization, cost control and to forecast future levels of operation. In that sense, cost accounting can also be considered as the language of finance used to interpret the value chain and develop cost profiles of activities, processes, products and services. Based on the type of industry, a relevant method of cost accounting can be used to report the actual cost performance (historical) and to forecast future performance (Horngren et al. 2009). Another area handled in the domain of cost accounting is the accounting of materials and resources. Materials and resource accounting (as a part of cost accounting) traces the temporal movement of the costs of materials and resources and generate inventory accounting records reflecting different types of transactions (consumption, returns, salvage and discards), from where periodic balances of physical inventories can be derived.

Cost accounting uses historical costs and follows specific rules of accumulation to generate inventory accounting. For example, inventory can follow a last in, first out (LIFO), first in, first out (FIFO) or any other disposition method, but each would need cost accounting to follow through in generating the temporal records of the materials' movement. Simultaneous accounting of physical quantities (inventory accounting) and value (inventory valuation) in cost accounting records is more granular than basic financial accounting and provides insights for effective inventory management. This includes tracking material wastes at the physical level, as the valuation rules transfer the accumulated costs of wastes to the final products to follow the accounting principle of conservative outlook (cost or realizable value, whichever is lower). Traditionally, costing methods adapted to the nature of the industry and the topology of manufacturing method, and accordingly developed job costing, batch costing (for engineering industries), process costing (for continuous or flow manufacturing), contract costing (for engineering, procurement and construction—or EPC—industries), transportation costing (for logistics), service and operation costing (for services) and other variations.

To detail some of these:

- *Job costing* is to determine, record and estimate the cost of individual jobs over a period of time and is particularly suitable for engineering jobs of short duration.
- *Batch costing* refers to estimating and recording costs for a manufacturing process that operates in batches. Batch quantities are dependent on the set-up and runtime of machines and the ability of the bottleneck resources to continue the throughput of the production line. Both job and batch costing are particularly useful in assembly manufacturing.
- *Process costing* relates to a continuous manufacturing environment, where the cost of outputs is distributed or averaged over large quantities of homogeneous products. This includes both continuous as well as repetitive manufacturing or production environments.

- *Operation costing* is another variety of process costing that is more suitable for defining the cost of services as against manufacturing, and is also known as *service costing*.
- Similarly, *transportation costing* is concerned with estimating, recording and accounting for trip costs of logistics services and computed to derive transportation costs per unit of distance travelled, per unit of transported materials.

There are a few different techniques of costing as well. To control costs in a repetitive manufacturing environment, for example, techniques like standard costing and budgetary control have been developed:

- *Standard costing* uses a pre-determined cost of a cost object within standard operational parameters and conditions. The difference between the standard and actual performance levels would accordingly generate variances that could be used for responsibility accounting and performance evaluations.
- On the other hand, budgetary analysis helps in establishing the targets, comparing it with actual performance and focusing on responsibility accounting for deviations (Horngren et al. 2009; Prasad 1977).

Costing and related techniques have proved useful for managing the information needs of organizations as well as in helping management with information in the areas of cost determination, cost analysis and cost control, not to mention life cycle costs of the products and services. Cost planning is an important element in prototype designing and planning, as it brings together the product and services designer, engineers and the management to validate prototypes that satisfy the necessary costing and engineering parameters to move from prototype to the production stage. The design-to-cost strategy offers a systematic review of how product costs are defined, designed, targeted and achieved. Cost determination is relevant in developing the cost structure of new as well as existing products and services (offerings). In the case of new products and services to be introduced, the target cost sets the threshold to move an idea from the conceptualization phase to the prototype and development phase, and later to commercialization.

With advancements in manufacturing intelligence and automation, costing has evolved techniques such as throughput costing, target costing and quality costing that define costs using a customer-centric approach (Horngren et al. 2009). Throughput costing helped with the organizational adoption of throughput manufacturing, which had its genesis in the theory of constraints (TOC) proposed by Goldratt in 1984 in his book *The Goal* (Goldratt and Cox 2016), which proposed driving efficiency in the value chain by identifying constraints that need to be removed from the process. This needed the costing process to follow through how value creation is defined or targeted, including to evaluate how it is different from the conventional costing method, and alter the cost view to connect costs to the value deliverance. For example, throughput value chain holds sales and marketing functions as direct to the value chain, so does throughput costing by accumulating these costs as direct to products or services, as compared to the traditional view that holds sales and marketing costs as overheads. The same view is extended to other costs as well, that is by relating costs to the business context and derive the ones needed to target constraints.

The same is the case with other costing techniques that were developed to be more customer-centric. Target costing, for example, developed to align the product cost planning for the design and prototypes of new products and services that would be better aligned to customer expectations, for conceptualizing product features that met or exceeded customer expectations, or even to match product pricing to customer expectations. Target costing is used in cost planning, during the design stage for products and services and for tailoring product features to meet customer expectations. This helps in predicting and controlling costs once the product moves from prototypes to production. In any case, target cost covers the cost elements that can be directly attributed to the features being included in designing products or services, whereas overheads or indirect cost are pooled to reflect common costs and added to the product costs as overheads. This includes distribution overheads, unless the product design also involves designing an innovative delivery method that won't share resources with the existing delivery channel. This also means looking at all the aspects of products and services from the perspective of design considerations, consumers and marketing expectations, engineering complexities, and using the relevant elements in defining the products or services. However, one can look at these various aspects from different perspectives, or while giving primacy to specific aspects:

- The *design view* is all about using design thinking and bringing together design elements to develop, define or conceptualize the product or service that meets the target costs. This could also include improvising design aspects to match user expectations that converge on the target cost. It can also extend to defining incremental improvements or group of features and the ascertaining costs corresponding to it.
- The *engineering view* is related to deconstructing the product or service in terms of its bills of materials, service components, routings, component-level improvements, and the materials and service characteristics for the product or service to match the target cost and bucketize deviations with a select set of sacrifices or gains. Feature-led product pricing is an example of this kind of exercise.
- The *marketing or consumer view* is about meeting customer expectations and developing features that the marketing team believes is close to the consumer expectations or consumers' willingness to pay (WTP). Here WTP translates to the bundle of product or service characteristics that consumers believe they are paying for and should be inherent to the product or service.

Quality costing is another technique that has evolved to guide the costing process to connect with the quality initiatives of firms and derive quality costs by accumulating costs of all the activities, components and processes that contribute towards improving the quality of products and services. Quality costs are related to the prevention, measurement and resolution of quality defects that products and services should be free from and allow firms to examine the reliability of their products and services and determine specific features that characterize these quality improvements. Accordingly, the costs associated with the quality control efforts are collected to derive the cost of quality conformance. This helps to relate the performance or the characteristics of products and services to their design criteria, and identify the costs associated with achieving or building these features or characteristics. Importantly, quality costs could be point costs (such as a cost of rectification), but their efforts might be spread over the useful life of a product or service, which brings in the time-delayed impacts of design considerations and obsolescence aspects due to technological innovations. For example, products or services with high technology-obsolescence rates would generally be carrying a high cost of quality so as to prolong the utility value of the product or service and keep them relevant for consumers in a changed environment as well. In comparison, ongoing costs can include the costs of prevention and diffusion, where quality considerations are built in while producing the product or service—for example, in the case of flow manufacturing, quality inspection is embedded within the production line and not only as an end-of-manufacturing operation. This 200 per cent quality test increases the cost of quality and prime cost of the product or service, though.

These new techniques of costing are well supported by a computerized environment of operations as well as by different enterprise resource planning (ERP) systems. ERPs can capture the costs along transactions and record these to develop cost sheets that can connect to not only financial and cost ledgers (Granlund and Malmi 2002), but also to the manufacturing or services environment and post-sales efforts. This includes product-, service- and costs-related analysis where data is pulled out and cross-linked against different contextual backgrounds to develop a cost profile that can offer insights in terms of temporality and life-cycle stages, and a better view of cost incurrence, profiling and causal deviations from the expected values.

MANAGERIAL ACCOUNTING AND DECISION-MAKING

Management (managerial) accounting is defined as the methods and techniques used by the management accountants, including information generated from the organizational and accounting records (including cost and financial accounting) to aid management with a scientific basis to the decision-making. The decisions supported by management accounting techniques could include operational (make versus buy, cost allocation, repair versus salvage, insource versus outsource), investment (invest versus outsource, fixed assets versus rental opportunities) and/or financing-related decisions (lease versus buy, invest versus sell). Management would need information to evaluate the evidence in all such cases before selecting any available option. In that sense, management accounting shifts the emphasis of costs from record-keeping to the generation and use of information for a scientific decision-making process. This would include simulating 'what-if' scenarios of different options and developing systemic perspectives for strategic decisions (Kaplan and Atkinson 2009). The theory and application of management accounting can be divided into three areas: (a) formulation of a decision problem, (b) collection of relevant information and (c) objective interpretation of information to help in better decision-making.

The core decision-making areas in general can cover those related to strategic management, performance management and risk management. To support the decision-making process in organizations, management accounting has methodologies in its arsenal such as activity-based costing (ABC), transfer pricing, capital budgeting and resource accounting. The performance measurement activities use techniques such as balance scorecards, budgetary analysis and profitability analysis (Horngren 2004), while risk management areas would need risk evaluation and mitigation framework. However, a detailed analysis of the standard discourse of these methodologies is not the purpose here, as these are extensively covered in standard texts. Instead, I seek to highlight in some detail the economic nature of cost and management accounting, being positioned within the economic realm of businesses and designed to use the common language of finance to translate every business activity into the common denominator of monetary valuation.

Even though management accounting uses different techniques to approach decision-making (examples might include a discounted cash flow for decision-making on capital investment, marginal costing for evaluating make versus buy decisions, and optimization levels for production activities), these methodologies basically follow cost–benefit analysis (CBA). CBA demands every decision variable be converted into a common denominator and uses financial value as a numeraire to compare aggregate costs with the benefits associated with a decision scenario, while discounting future costs and returns to the present value. This is not a problem of selecting a particular right choice. It is a systemic concern with expressing decision problems against a contextual background. At the same time, the construction of the contextual background is dependent on our current knowledge and how well we have been able to abstract it.

The historical nature of cost accounting and charging overheads to cost centres has its own share of criticism. It forces decision makers to rely on historical data and extrapolate available information to build future projecttions. For example, experts have argued that the uniform rate of charging overheads to cost objects is not causal in nature and does not translate into effective cost control. Neither does it offer a way to control overheads. This has led to the development of new costing methodologies such as ABC, which has subsequently evolved into activity-based management (ABM), where the movement of costs is traced along the chain of activities and can be better related to the products or processes. In this case, costs are pooled based on activities and later distributed based on the cost drivers. This method allows a more causal approach to controlling costs of activities. However, at times this would not help connecting, for example, how value addition offered to customers through products and services match customer expectations, and how to target building specific features that customers would love to experience. This brings us to the question the relevancy of management accounting in different decision-making situations. The tools and techniques of management accounting are not prescription-based but based on the practitioners' knowledge, the techniques' relevance to the situation or context, and the anticipated outcomes (van der Meer-Kooistra and Vosselman 2012). Accordingly, practitioners are expected to be knowledgeable in the latest techniques and their relevance to a given situation, not to mention the objectives that the problem statement intends to achieve.

In contrast, although advanced techniques of lean accounting (in a lean manufacturing environment) and throughput accounting (along the value chain) have been developed to capture the redundancies within the customer value creation process, they have not been designed to ameliorate the negative environment and social impacts of products or services either (e.g., improve dematerialization, reuse products, reclaim the ones nearing end-of-life, and so forth). Along similar lines as CBA, cost–volume–profit (CVP) analysis and marginal cost analysis mirror corresponding economic methods for decision-making within organizations. CBA is used to compare the present value of the future benefits of any strategic decision and the associated costs, accumulated at the present value at a specific discounting rate. Rooted in the economic principle of the present value being higher than the future returns, the discounting factors are adjusted to transfer the costs and benefits to the present value. Similarly, marginal analysis is relevant to the economic theory of the marginal cost of production, as it provides organizations with information on the incremental costs that they have to incur to increase the production of a given product by one unit, assuming no change in capacity. Similarly, CVP maps the movement of cost versus volume and profit to enable firms to economize on the volume of their activities (Bebbington, Brown and Frame 2007).

Over the years, management accounting has evolved into strategic management accounting (SMA) to support the management in strategic management-related efforts. Nixon and Burns (2012a), in their editorial in the journal *Management Accounting Research*, which published a special issue on strategic management and SMA, pointed to the management's need to respond to environmental challenges and change business paradigms quickly and efficiently—this needs SMA to advance tools and techniques that can connect to changing paradigms and shape the management response. Nixon and Burns (2012b) offer evidence that SMA techniques have not been widely adopted. Yet, while they acknowledge the decline of SMA, they also agree to sustained growth of SMA tools, models, and concepts that seems to suggest its growth along strategy formulation. However, low recognition of the term 'SMA' and a lack of consensus on the definition of SMA remain a cause of concern.

From the 1990s, the problems of strategic management were seen more as systemic concerns. Johnson and Kaplan (1987, p. xix) stated that 'management accounting systems are not providing useful timely information for the process control, product costing, and performance evaluation activities of managers'. Although the strategic cost management framework was comprehensive and conceptually robust, it was not widely adopted. Shank expressed disappointment that strategic cost management was mostly 'evolving outside the purview of the accounting profession' (Shank 2006). However, there is no clear-cut timeline that puts management accounting and SMA into two distinct time periods. Evidence suggests that how SMA techniques and processes diffused into general practice within organizations remained less visible (Langfield-Smith 2008). Moreover, the role of information technology in augmenting management accounting tools and techniques remained much less evolved but hopefully will grow in future.

INFORMATION BEYOND NUMBERS AND INTERPRETIVE ELEMENTS IN DECISION-MAKING

Business management has substantially relied on the development of economic methods where practitioners and academicians abstracted economic theories to fit the boundaries of business needs. This has resulted in the organizational practices to simply deny the complexities attached to the processes and their interaction with the organizational structure as if these are irrelevant. Moreover, the simplicity of using few variables to structure problems annul the need to expose the complexities to the decision-makers or to improvise the problem situations. Our habit of selective abstraction and dealing with linearity has not only become counter-productive in the long run, not to mention the tools and methodologies that were denied the chance to gain strength by handling difficult problem situations with other business areas—for example, financial management, general management and organizational behaviour. To elaborate, financial management is significantly diffused with accounting and has brought in the need for accountants to depend more on the methods and measures that have been borrowed from economics and mathematics and contextualized for decision-making in commercial organizations. The methods of evaluating choices based on financial stakes (like discounted cash flows and payback period calculations) lead accounting experts to strip the surrounding information of choices for decision-makers, abstracting and defining the variables necessary to fit the choices. Whether a select set of elements present the problem adequately to make informed decisions is one side of the story, discounting future earnings assuming a higher profit upfront is more beneficial for companies instead of delaying returns for future is another one, bringing in narrow-focussed approach of econometrics to management areas. Incidentally, this is contrary to the entrepreneurial spirit, opposite of taking risks for better future by discounting the gains of today, so as to ride the opportunity waves and shape the competitive landscape of tomorrow. How can accounting techniques evaluate these opportunity losses?

From the preceding discussion, we can summarize that accounting methods are based on a conservative outlook, and are not innovative enough to be changing and evolving vastly with time. On the contrary, professional management in companies is expected to be focused on innovating products and services that delight customers and on the competitive landscape, and not on numbers beyond a point (hopefully!). I have an interesting example from my own experience to share. I have been in technology consulting for multiple businesses of General Electric (GE) within the last decade, focusing mostly on process automation and improvements. I learnt advance business fundamentals from GE that any b-school curriculum is yet to theorize, including theoretical foundations of competition and sharing stage with competitors through coopetition (bringing competition and cooperation together), but what caught my attention was also the enforcement of fiscal discipline and practices of measurement that translated every improvement to a corresponding dollar value, emphasizing the six sigma that is well disseminated in literature. As of 2018, GE has been removed from the Dow Jones industrial index and the industrial behemoth is struggling to be at the zenith of what it is known for: innovation. We are well aware of the research- and innovations-led achievements in GE for over a century, but did it suffer from excessive emphasis on measurements and a quantitative outlook that became counterproductive to the entrepreneurial spirit of innovation that GE was known for? I would leave this riddle in the able hands of researchers in strategic management.

I would, however, question the overemphasis on the economic sciences and quantitative methods and their role in the accounting sciences, which is a two-fold issue: (a) any attempt to isolate business risk as specific to business only resonates with the belief system that decisions created within the business boundary do not impact anything outside the firm and (b) it is arrogant to persist with the argument that the languages and tools used in this process are evolved enough to interpret and translate everything that we need to know and understand in order to let us act and react judiciously. In addition, we are also facing the interpretive challenge of encapsulating these as discrete choices expressed within a specific set of variables, which cannot effectively and correctly represent the entire problem situation. This is not a failure of any specific method or technique but a systemic misreading of a situation that is derived out of oversimplification and a collective (mis)understanding on treating a problem situation in a certain way. This belief system and our resulting orientation of confining concerns within the specific framework that we connect with is much better as compared to the issues themselves, and we can somehow force-fit the situation that requires information specificity, discarding a part or more of what cannot be handled, helping ourselves remain oblivious to it in the long run. This mindset has been referred to as the 'McNamara fallacy' in literature. Daniel Yankelovich (1972) used this expression in the book *Corporate Priorities: A Continuing Study of the New Demands of Business* for the first time:

> The first step is to measure whatever can be measured. This is OK as far as it goes. The second step is to disregard that which can't be easily measured or give it an arbitrary quantitative value. This is artificial and misleading. The third step is to presume that what can't be measured easily really isn't important. This is blindness.
>
> The fourth step is to say that what can't be easily measured doesn't exist. This is suicide.

Economics
A Rational Argument to Ignore Environment

INTRODUCTION

Industries and business firms are important constituents of human societies, continuously influencing and shaping the way human societies have been interacting with nature while fulfilling human needs. Societies might choose to accept the way organizations interact with nature in their bids to justify economic compulsions, which is an acceptable form of existence for economic organizations today, but we cannot annul the externalities that the firms produce, nor control how nature will counter the impacts of anthropogenic changes. In this process, waste and emissions are considered as 'acceptable' byproducts of economic interventions that add to societal responsibilities, now or in the future. Even if we believe this is an acceptable form of existence and we choose to live with it, we would have to be open to the idea that this acceptability has certain inherent costs—conceivable or otherwise, now or in the future. Along with debating what these could be, this chapter has covered a normative discourse on the established theories of cost and management accounting, highlighting their limitations in how the prevailing accounting practices are inadvertently shaping the business decisions to ignore environmental impacts altogether. Also, this deliberation is relevant in appreciating the difference between the mathematical construct of accounting and the meanings it offers in form of symbols and information, and whether the process can be leveraged to work towards the social and environmental commitments of organizations in future as well. Some of these vexed issues are explored in this chapter.

ENVIRONMENTAL VIEW OF ORGANIZATIONAL ACTIVITIES

Business organizations are agents of economic growth, engaged in large-scale operational activities such as mining, manufacturing, processing, marketing and distributing of goods and services to be consumed by society. However, along with the goods and services, organizational activities also generate unwanted by-products such as waste and emissions (environmental impacts), which are considered an inevitable outcome of the economic activities. The Industrial Revolution is unmatched as a human endeavour of such scale that it could redefine how humans choose to express their collective intent and redefine the ways in which we interact with nature. The collective efforts of humans through the machine culture of business entities have been shaped to meet societal demands for goods and services and, in turn, pushed the organizations to evolve to satisfy, (re)structure, improve and develop the ways and means to increase our material wants. While doing so, trails of progress—or scars—have been drawn on the earth's surface. Societies have chosen to live with these side-effects as an inevitable outcome of the technological limitations of the conversion process, inherent to the developmental pathways. For example, waste that gets discarded during the conversion process of raw materials into finished products and services (material waste, waste water and emissions) or generated before, during and after the consumption of products and services (e.g., packaging, parts discarded during maintenance and end-of-life disposal) ultimately ends up in the public pool. The unintended effects are the environmental impacts and externalities are by-products of the process, and inevitable as per economic theories, not bargained or paid for by the consumers and society, neither controlled by the market forces or its invisible hands (McLaughlin, 1993).

Economic theories have considered externalities as market failures because the 'invisible hands' of the market have failed to control the flow of unintended goods and services. Neoclassical economics has treated pollution and all forms of waste as market failures or negative externalities, and its subsequent treatment has been a never-ending debate between different schools of economics. Neoclassical economists have chosen to avoid externalities by limiting these to be spillovers in mainstream theories due to the difficulties associated with rearranging the resources, and the comparatively miniscule nature of these externalities, they argue, does not shift the (theoretical) market equilibrium dramatically in the absence of any treatment (Dahlman 1979). On the other hand, new institutional economists are of the view that such implications can be handled through contractual agreements between impacted parties wherein the associated transaction costs might lead to a suboptimal arrangement—a solution rooted in the incompleteness of contracts between parties—á la Coase theorem. In other words, negative externalities can be viewed as an opportunity to generate optimal contracting arrangements, costs of remediating which (it is theorized) can be salvaged through private contracts and rearranged transaction costs (per transaction cost theory), even though the exceptions to the Coase theorem remind us about the inability of a private arrangement between parties to reduce the impacts of the publicly disposed private waste (for example, smoke releasing different pollutants into the air).

Since waste and emissions are not always environmentally benign in nature, their subsequent handling and disposal activities would need public resources (infrastructure), not to mention the costs associated with marginalizing or reducing their impact on public goods (ambient air, clean water bodies, uncultivated land, forests, etc.). The cost of remediation or neutralization of waste and contaminants is neither understood completely (other than the occasional environmental accidents and runaway effects), nor are covered by the firms through fees paid for disposal activities, which is resulting in social overheads in the form of present and future costs that are (or will be) incurred by society to mitigate the negative impacts of these goods. By equating negative externalities to the social costs, free-market economies have allowed the market forces to *incentivize* the correction of environmental wrongs (Block 1998) and develop solutions such as market-based instruments and the command-and-control approach (Pirard 2012). Market-based solutions are a way to incentivize firms to discover ways and means to reduce the production of unwanted by-products, with a hope that a market incentive would prove to be a motivation for firms to behave in ways that would not require a deterrent. However, since the environmental goods (air, water and soil) are outside the market valuation, this suboptimal arrangement results in private gains in the short term at a cost to the public and sustainability losses in the long run.

OWNERSHIP OF WASTES AND THE TRAGEDY OF THE COMMONS

Besides the moral dilemma that undermines 'shared responsibility' in favour of the right to waste or pollute (Gayer and Horowitz 2005), the present discussion also intends to highlight the obscure nature of the ownership of waste and the absence of a legal framework to ascertain it, which has helped the industries to shift the private burden of externalities to the public domain. Waste in any form (such as industrial fluids, waste water, air emissions, and nuclear and biohazards) is a non-saleable, valueless output for the firms, meant to be discarded. However, the moot point is whose waste is it anyway? Following the economic processes, firms purchase materials and/or services from the market. The purchased materials and services carry ownership rights along with the goods and services that are transferred from sellers to buyers through the commercial transaction and duly supported by the legal frameworks upholding such rights. The same is the case for manufactured goods, which carry ownership rights that change hands depending on the terms and conditions of sale. However, economic studies on waste have stopped short of defining ownership of waste upon its transfer from industries and private arena to waste-management facilities and/or the public pool (Bose and Blore 1993; Chaturvedi 2003; Pongrácz and Pohjola 2004), except for radioactive wastes.

Irrespective of whether the waste disposal mechanisms are privately owned (as in developed economies) or owned by the local administration (as in the case of developing economies), the ownership of waste has not always been covered under express or implied contracts upon its transfer, except for when the legislation relates to an unequivocal right to benefit from the use of discards by treating and using it by any means and in any manner that a firm deems fit, which is also associated with risks, harms and dangers that get transferred to the public space, outweighing the gains assumed while proposing such legislation, if it even exists. On the other hand, in the absence of any express legislation to this effect, the ownership of waste is retained by the firms discarding it, although it can be characterized as a property with relinquished or abandoned ownership rights but yet to become the exclusive responsibility of society to deal with, unless legally provisioned for (Pongrácz and Pohjola 2004). Further, in the case of radioactive wastes, the provision for transfer and disposal of high- and low-level radioactive wastes (HLRW and LLRW) are covered through special legislations in countries like Sweden, Canada and the US, where private firms are allowed to participate in civil nuclear programmes (Sjöberg and Drottz-Sjöberg 2009). In other countries, where the nuclear facilities are under government control, the waste-disposal activities of the facilities are handled by specialized agencies. This is based on the acknowledgement by the stakeholders (the government and society) of the need to demarcate the transfer of responsibility for nuclear wastes in the post-production period.

Although the ownership of wastes and legal jurisprudence to enforce ownership rights outside organizational boundaries are inadequately studied, if legislations can exist to reclaim specific classes of waste, its absence in other cases does not translate to an automatic transfer of ownership from private to the public domain. Yet an opposing view suggests that the unique nature of radioactive wastes and its harmful effects can be traced back to the source, so such an arrangement makes sense. Still, unless the waste is proven to be environmentally benign and its impacts on the natural environment upon entering in the public pools are neutralized, it cannot become societal property because of failure to link it back to the source. The same is the case with emissions. This leads to the question that if waste does not rightfully belong to the society, why should the harmful impacts and costs associated with mitigating their impact be borne by the society? In principle, should this not be the responsibility of its rightful owners, the producers? Moreover, if these are abandoned goods that ultimately end with society having to take care of them, should the cost (rent seeking?) not be recovered from the set of entities that produced it in the first place, that is, the industries? Within the underexplored crossover of legal and environmental boundaries, the question remains.

In view of the inevitability of the negative externalities and the market's failure to regulate them, Pigouvian taxes (or a command-and-control approach) have been proposed in the economic theories to regulate the supply of externalities. Other than that, market-based instruments have also been proposed in the welfare theories to be used to develop permit or quota systems with the intent to create an artificial shortage of permits in the market and incentivize the participating entities to lower the level of externalities. However, these mechanisms would need overwhelming support from the legal structure in defining the terms of a cap-and-trade system and in developing level playing fields for the participants. Contrary to expectation, an asymmetry in taxing (or levying duties on) the polluting companies does not benefit the citizens being affected in the process, not to mention that the long-term nature of the externalities runs counter to the short-term economic measures (Dascalu et al. 2008).

WASTE AND EXTERNALITIES: SYSTEMIC PERSPECTIVES

If we view an organization as a system, its interaction with the environment can be considered in several ways. For example, the organization can be abstracted as an open or a closed system, depending on the perspective adopted. From the perspective of the flow of materials and resources, we can consider an organization to be a closed system which interacts with the environment to receive inputs (materials, energy, water and other resources) and converts them into products and services that are exchanged for economic benefits, while the discards are released back to the environment. Once expelled, the waste ceases to be the responsibility of the firms and its subsequent impacts are not accounted for by the firm. However, going by the thermodynamic flow of energy (available useful energy), the environment works as the sink in such cases and would need equivalent energy or more to deconstruct it to a neutral form. This would add to the energy requirements in nature, meaning extra work to be done by nature to make the waste productive for the ecosphere or leave it unattended (e.g., construction, demolition and radioactive waste would take years together to be of any further use). Even though it may look like that inertness of waste is an acceptable quality criterion to let nature remain unresponsive towards it, this would soon make the entire world a dumpyard, going by the rate of industrial consumption and the appetite of societies to become affluent, not to mention that it would be against the ethos of a living system, where every part evolves in tandem with others and participates to define the characteristics of its future. Creating dead zones in isolation might not push the ecosphere to change in the short term, but long-term impacts are still not within the realm of scientific advancements.

In comparison, if waste or its constituents are directly injected into a benign natural environment, as in the case of carbon dioxide (CO_2) and other greenhouse gases (GHGs), waste flow would increase their concentration in nature (depending upon the flow and absorption rate of the compounds) and introduce changes that might be difficult to predict for now. For this effect to be predictable, scientists might have to wait for the concentration level of the elements to breach a certain threshold, beyond which the system won't adapt to the increasing concentration of elements and will modify its behaviours permanently. This is why the impact of a 1.5–2 degree rise in temperature is being talked about as a threshold (Carter et al. 2007), but that's only the best guess we have based on the accumulated scientific knowledge of what we know about the environment as a living system. Moreover, it is still not accurately known whether there is any recourse available, mainly because we are yet to scientifically establish the complex inter- and intra-connectivity of material and energy flows within the complex adaptive system and to accurately predict its future transformation and reversibility. From the economic perspective alone, waste is an outcome of the economic system that the market forces will not care for unless accompanied by certain value credentials. Human society has chosen to ignore the warning signs that the one-sided perspective of economic theories has thrown up and needs to improve the concept of value to reflect overall contributions.

While the process of transferring private waste to the public arena has led to private gains at the expense of public goods (loss of public welfare), the continued carelessness of economic agents to ignore externalities have led to environmental degradations, unsustainable use of resources, increasing levels of waste and a suboptimal life cycle of materials. While industries are generating private profits by paying for what the market mechanism believes to be adequate for disposal activities, waste and emissions are being transferred to the public pool and result in societal responsibilities. In the absence of legal recourse, society is unable to seek rent and build a corpus to install a mechanism that can work towards reducing the damages that wastes would eventually lead to, leaving the ecosphere in a less than desirable state for the future generations. Accordingly, pursuing economic progress has resulted in an uncaring attitude towards consumption of materials and resources which follow a linear flow of cradle-to-waste, thereby impacting the regenerative capacity of the earth and robbing future generations of their share of the natural capital, not to mention pushing the ecosystem to the tipping point, beyond which the present level of support for life might get permanently altered. Although businesses are rising to the wake-up call and seeking ways to change their behaviour, the efforts are muted because of upholding economic considerations while avoiding, if possible, legal and social sanctions.

Current accounting practices are scaling new economic heights but without warning firms about the harmful impacts they produce. The essence of this argument lies in the economic rationality of the firms, which has been the predominant paradigm per which to view its contributions. Even when shareholders were replaced with the term 'stakeholders', the need emerged for accounting to support the accountability of organizations' management towards a larger group, an effort which has been caught up in the methodological shortcomings. As a result, organizational practices are not being scrutinized (in a systemic manner) for what they ultimately leave for nature and society to deal with. Like any other activity within an organization, accounting practices have been active in shaping the behaviour of firms by translating business performances into discrete sets of information that characterize its investability, but not beyond economic sustenance. Similar is the case with traditional cost and management accounting, which has certainly produced some good methods and tools to analyse costs and improve informational support to management, leading to better utilization of materials, funds and resources, but are hardly capable of evaluating environmental considerations as (a) organizations are yet to learn how to include environmental considerations as part of a decision-making process, (b) the traditional techniques were not devised to capture data on environmental aspects other than the quantity of waste, (c) the methods to quantify environmental aspects are yet to be standardized, (d) the methods and techniques of cost and management accounting are based on historical costs and not on environmental aspects and (e) the lack of uniformity in interpreting environmental impact reduces its usefulness in decision-making.

SUSTAINABILITY AND OUR COMMON FUTURE

Sustainability, reflecting the philosophy of preserving intra- and intergenerational equity, is not to be confused with sustainable development or equating the two (Gray and Bebbington 2001) due to, is a philosophical arguments over the present level of consumption of materials and resources, which is believed to be robbing the earth of its natural carrying capacity and its ability to satisfy the needs of future generations. This relates to the insatiable hunger of the world economy in excavating minerals and fossil fuels to drive economic growth, churning out goods and services at a breakneck speed and forcing hyper-consumption levels. While economic growth may have regional variations due to the differential rates of industrial progress and the growth models adopted, overall the process is responsible for placing the environment under continuous stress to counter the negative impacts of anthropogenic events and maintain sustenance.

Scientists and ecologists fear that the increasing levels of wastes and emissions are exceeding the absorption rate for nature to maintain the present level of sustenance and we could soon reach a stage that might result in a permanent shift in behaviour of the ecosystem. The changes within the ecosystem, if permanent, could create effects that are not very well understood within the current body of knowledge (Daly 2005). Gowdy and Hall (2010) have offered a comprehensive assessment of the philosophy of economics that views the world and its interactions using the unidimensional approach of Walrasian economics, with a closed cycle of interaction between firms and consumers influencing policy-setting parameters to apply a cycle of prices and a Pareto improvement to every problem are in contravention to the laws of thermodynamics and biology. Instead of considering anthropocentric values, if the goods of the same type are not considered substitutable and carry value for the entire ecosystem, 'value' that is not known or understood within the economic parameters, that outlook impacts the entire ecosystem. This also questions if discounting is the only way of bringing in future impacts of policy choices, which translates to the immediate need of discarding the current core of the economic model and replacing it with a new and a better theory.

Ulrich (2010) has decomposed the ideological basis of sustainability and questions the relevance and the means used to achieve it, which in his opinion would never come out of the present mess. Questioning the economic outlook and political mindset required to drive such a change, he argues that a great deal of social and ethical unrest is foreseen that would push the world towards even more unsustainable trends. He would rather let the national economies join hands together in creating a supranational stage that works towards a common theme, away from selfish national identities. Its goals should be guided to arrive at a more mature perspective of social development that measures the human standard of living by relying on life-enriching values instead of the level of material possessions. Considering sustainability as an integrated element within the social, political, environmental and economic spheres, we need to look for environmental effectiveness of pursued policies, at least to begin with, rather than looking for eco-efficiency.

Another way to look at it is by taking a cue from the recent economic slowdown around the world. If the economy could be considered at three levels—financial (assimilation of flow of money, values and associated transactions), productive (involving the manufacture and creation of goods and services) and real (source of energy and materials)—the development of Keynesian economics as the engine driving long-term growth and survival might not be a sustainable one (Alier 2009). For example, replacing one source of energy (fossil fuels) with another (renewable sources) does not account for depleting resource levels (degenerative ones), whereas the degrowth or negative growth theory could help in reducing the speed of consuming materials and resources. The recent negative growth of developed economies (the recession of 2008–2012) could be taken as an opportunity to think about the real economy where sustainability and local growth, employment and conservation of resources go hand in hand. However, this would require new methodologies and for social organizations to manage the uncertainties and ethical complexities.

Although the subject of sustainability is explored in this text as a concept that serves as a background against which to place the relevance of other ideas being discussed here, the point remains that sustainability as a philosophical approach is about improving the future state of the earth, and the concept is valid even when the so-called future state is not defined as an end state, but remains more of a process. Accordingly, sustainability has been discussed here along with economic, environmental and social aspects of development, and unless the environmental view of anthropocentric activities can be analysed differently than it is being done today, our focus won't shift from economic well-being, where our language and expressions are yet to be backed by scientific evidence beyond the sense of altruism that sustainability evokes. The current worldview to support or negate any activity, action and decision is solely grounded within the economic rationality, including to analyse—for example, if a common grazing ground for animal herds will always be susceptible to overgrazing and loss of biodiversity unless an entry fee is placed as a market operation for equalization, which would be a controversial idea if analysed from the point of view of nomads and their dependence on nature for common goods. In contrast, an environmental view lets participating entities view things from the ecological perspective—for example, seeing the common grazing ground as a property of the group that needs to be looked after by all who share the bounty. This view is not based on the maximization of individual wealth but on an equitable distribution of benefits from joint resources. Social behaviour on other occasions helps us reflect on the altruistic nature of individuals and groups (a topic inadequately studied), where the economic rationale is defied in favour of contributing to the collective wellness of the society. However, we can be certain that our current understanding of environmental interactions is not based on a complete understanding of how Nature acts and reacts to anthropocentric activities. The point is, more than economic theories, it is the economic thinking that is driving how we look at things and take decisions. Within this paradigm, ideas like 'inclusive growth' and 'care for nature' fail to find expression simply because the language of economics is based on improving wellness (utility) through material distribution (based on needs), which unfortunately signals abandoning altruistic thinking. Accordingly, further research is warranted to improve our knowledge of human–nature interactions, as well as to develop a language that can express these interactions from ecological, and not economic, perspectives. Between the extremes of minimal change to business versus minimal disruption to the earth's carrying capacity, the question of how (un)sustainability of human activities is to be viewed against the backdrop of everything else, remains unanswered and forms a part of the enquiry in the next section.

Environment and Accounting Theories: Contemporary Advances

Environmental Considerations and Conventional Accounting Theories

INTRODUCTION

Accounting and other financial constructs are tools available to modern management that aid in the decision-making process by bringing in a scientific approach to analyse and construct the contextual information surrounding these decisions. The rationality of any decision can be preserved and traced back to the facts that are extracted from verifiable information generated by these tools, by leveraging data and information that were recorded by following accepted principles of business conduct, including the corresponding financial implications. This includes information and reports generated by the accounting systems and use of the information by different methodologies from managerial accounting systems. Since it is normal to accept a monetary justification for pragmatic and forensic evaluation of economic activities, the current form of accounting information is enclosed within the economic sphere of business, and although our widespread and continued dependency on it can also be construed as this being a universal language of choice, it cannot capture and express all the choices available beyond what can be translated economically. In this chapter, we explore why environmental considerations have always been a missing aspect of organizational behaviour, which severely limits the usability of information to effectively handle the contemporary challenges. A common language, if available, would have enabled firms to deal with rational as well as emotional choices—improving the quality of life of employees is one that a firm might find interesting to pursue, or finding a way to preserve something that is universally considered priceless (for example, the call of a songbird—and allowed for a plurality of commitments and perspectives, reflecting shared responsibilities and a multidisciplinary approach that characterizes ecological wisdom (Funtowicz and Ravetz 1994).

PREVAILING ACCOUNTING THEORIES AND THE STAKEHOLDERS' RIGHT TO INFORMATION

While accounting theories have evolved over time, their theoretical underpinnings have always guided the science and art of meeting the information needs of primary (owners and shareholders) and other stakeholders, including creditors, lenders, bankers, management and the public in general. The fiduciary duty of the management as trustees of a firm who protect the interests of the shareholders and owners has been an accepted premise within the theory of the firm (Chapter 2). As accounting theories are grounded in the information needs of stakeholders in relation to the economic affairs of the firms, at a minimum, this viewpoint corroborates for the foundational elements of accounting and their traditional proximity to the economic paradigm. For example, even when the theories debated whether the accounting constructs should be based on prices (historical costs) or (economic) values, the relevant needs of the owners and shareholders were examined to promote a universal norm of choice. As Paton (1922) insisted, 'it is the function of accounting to record values, classify values, and present value data in such a fashion that the owners and their representatives may utilize wisely the capital at their disposal.' With time, the expansion of accounting frameworks broadened the operational coverage expected of accounting, but the generalized approach still remained relevant to the stakeholders' needs. In a futuristic explanation of the role of accounting information almost a century ago, Fisher (1930) predicted that 'as interest in accountancy grows,... the public will demand that accountants furnish the information needed by investors and the consumers, as well as that required by proprietor.' Even though the stakeholders' theory

Environmental Considerations and Conventional Accounting Theories

stretched the perimeters of organizational accountability, businesses and market expectations remained focused on the information needs of investors, who are exposed to the risks of economic turbulence in the marketplace.

With this background, it is not difficult to relate to the normative view of accounting theories theorizing how an accounting framework, along with relevant standards, rules and procedures, would offer norms and guidelines to the practitioners (accountants) to interpret business events by translating the real-world business activities into data and information that reflect how firms behaved in the past. Throughout the transactional history of firms, interpreted as per the norms, recorded events in the books of accounts represented a specific instance of reality that allowed users of financial information to reconstruct a view of the firm's history and extrapolate to predict its future performance.

- In this regard, let us review the some of the guiding principles of normative accounting, and how they bring in linguistic specificity into practice. The accounting *principle of conservatism* guides the accounting practitioners to be prudent in anticipating possible future liabilities that the firms might have incurred through their business conduct and record these as well, while remaining restrictive about reporting and accounting future revenue streams unless these have passed the test of significant assurance. This allows users of accounting information to explore whether a firm has accumulated a potentially significant liability to be concerned about in the short as well as in the long-term.
- Similarly, the *principle of historical costs* guides how organizations have been managing their economic imperatives while ignoring the time effect of money. Although recalibrating a certain class of assets and liabilities to reflect the current state of their market position is allowed (by following market valuation, as against historical costs), accounting has traditionally suffered in working with other forms of costs, such as economic or indexed costs, the use of which could result in introducing certain elements of uncertainty due to these cost compositions being notional or discretionary in nature. Use of historical costs absolves the accounting sciences from validating the

numéraire and instead honours the value derived from a transaction at a point of time.
- The *going concern principle* lets the usability of accounting artefacts and narratives stay relevant beyond the current financial year and, accordingly, conveys that accountants need not be cognizant of specific reasons to negate the future continuity of a business.

These principles support how accounting frameworks have formulated guiding standards to be used by practitioners to interpret and record the economic affairs of firms over time and to generate artefacts that have played the dual role of being an information agent as well as an enabler of economic choices for audiences and investors. Advancements in the standards and principles are also shaped by the demands for information and focused views from the market, and need not necessarily cover aspects of organizational performances that are important in their own rights but not so from market perspective (for example organizational contributions to the regional, social, and environmental imbalances). However, in terms of practice, financial accounting provided insights through historical records to depict how funds secured by the management and available for disposal have been used within a given period. This includes reporting information on the economic performance of firms for owners, shareholders and external stakeholders (creditors, investors, bankers and regulators) to evaluate risks associated with its market worth and future investability as a going concern.

In comparison, data and information generated by cost accounting records help management with operational efficiency and in achieving long-term objectives of the firm. Similarly, management (managerial) accounting supports businesses and decision makers through the decision-making process by offering methods and techniques that leverage information derived from historical data. While managerial accounting can afford to be innovative in places and as a part of internal decision-making processes (as compared to financial accounting), its theoretical underpinnings lie within the boundaries of value creation and hence it operates to fulfil that intent. In contrast, tax accounting allows businesses to remain compliant to the tax regulations of the region within which they are operating, with accountants focusing on adherence to the norms rather than the appearance of the artefacts. The objective here is not to examine the inherent difference between these frameworks, but to acknowledge that these frameworks represent separate but equally important systems of examining business transactions in different ways, with each one of these based on a viewpoint that is fundamental to what the framework preserves and protects.

Different accounting frameworks provide necessary support (in the form of standards, procedures and principles) to practitioners (accountants). enabling them to analyse business transactions in a manner that is grounded within their respective viewpoint and produce data and information in accordance with the intent of the framework. At the same time, they need to observe the principles of transparency (being open to third-party scrutiny and correspondence of balances), materiality (material in nature and content and not speculative or limited in contingency), faithful representation (matching the audit trail that traces back to the business event) and neutrality (not being influenced by the individual biases of practitioners). So, while transforming organizational activities into historical records, accounting language promotes selective acuteness to view organizational activities and interpret these by following the principles that the framework provides and to remain relevant to information needs. This interdependency (information needs and information flow) is bidirectional and has evolved over time, as stakeholders influence different accounting frameworks through their demands for information. At the same time, information and artefacts generated and disseminated provide inputs and shape these needs, thus forming a continuous requirement-to-fulfilment cycle (see Figure 6.1).

In contrast to accounting, when we refer to environmental sustainability as a responsibility of businesses, we are not referring to their economic existence but relating how the economic choices of firms are impacting the ambient environment and public goods, which the accounting frameworks have not been designed to capture. Nor are the frameworks cognizant of the impacts that business activities would

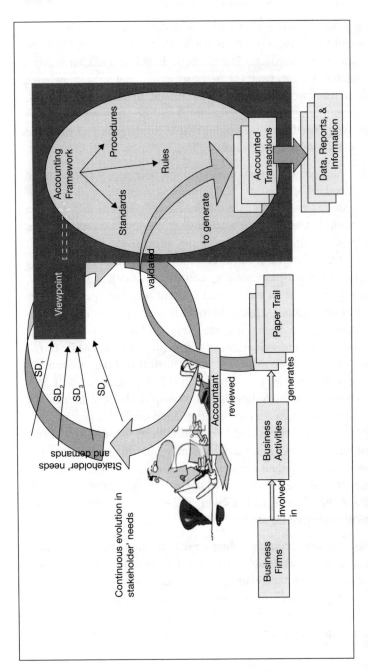

Figure 6.1 *Accounts information life cycle*

Source: Created by the Author.

Note: SD here means stakeholder demands

have on others, for example, on the environment or on society (quantified or otherwise), except in case of the legally enforceable claims of third parties, including tort claims, subject to the legal provisions of the region within which the business is operating and to account for certain taxes and liabilities in accordance with the environmental laws of the land. So, while businesses might have started embracing the principles of sustainability and social responsibility at some level and although the quantum and manner in which these enactments have been operating remain arguable and needs further deliberation, the accounting frameworks otherwise would keep ignoring them. The point relevant to this discussion is that although we perceive accounting practices to be restrictive and less caring towards the environment and of sustainability for now—and therefore, we believe that they are not living up to the contemporary challenges, as critical theorists have argued in the last two decades—we need to admit that the existing accounting frameworks are not designed to go beyond the economic choices of firms and stretch beyond the already defined objectives of the existing frameworks.

This is where the role of accounting practitioners comes into the picture, where the conduct of the accounting profession requires practitioners to remain faithful to the defined norms of the framework, and *not* to be innovative and offer or generate data and information beyond what has been prescribed as the norms within the framework. Before we start questioning the intent of accountants and their deliberate short-sightedness in remaining confined within the chosen boundary, let us consider the subsequent use of the information, including difficulties for the recipients in using it for decision-making or any other purpose if a rider is attached to the information artefacts that says 'the practitioners have chosen to be innovative during information generation process', only to find substantial erosion of trust and reliance. Here, it is worth mentioning that accounting practice is grounded in the ability of the practitioners to uphold the norms of the framework within which they are expected to operate, even at the cost of being innovative. A second layer to ascertain the objectivity of the process is the audit and assurance process, which validates the interpretations to ascertain they are in line with the GAAP and guarantees the neutrality of the artefacts or information contents being disseminated. As a result, it is not difficult to comprehend that the accountants would be constrained to interpret business activities in a way which is beyond the norms and/or at a considerable distance from how the frameworks have guided to interpret reality, including, for example, accounting for and report certain business activities might be potentially discouraging and harmful for the environment. A case in point is that of biohazards from specific industries that have significant impact on the ambient environment (nuclear medicine, nuclear power plants, extractive and forestry industries, and so on).

If the freedom to interpret business events is made available to the accountants, the accounting artefacts will become open and subjective to the interpretations of practitioners. While the apparent distance of accounting artefacts from other aspects of business could be held as the inability of the accounting framework to adapt to the changing demands of the times, we must admit this is how the frameworks have been designed to operate. Accordingly, it would be unfair (here I am in disagreement with the critical theorists) to expect that the accounting sciences and practice would interpret business activities to develop, support or forecast the (un)sustainability-savvy behaviour of firms, especially when this has neither been the intent of the underlying framework, nor the lingua franca of the mechanism through which the business events are being interpreted or recorded in the transactional history. While this also means allowing an overriding pass to the accounting sciences to remain blind to the other facets of organizational existence that ultimately contributes to not improving social or environmental ethos of firms, which if remains outstanding for long, would lead management and stakeholders to be less dependent on the accounting function.

But where does this lead us to? Should we agree that the accounting frameworks cannot adapt to the concerns raised previously or investigate why these demands cannot be addressed within the current form of the accounting sciences?

Researchers and thinkers have agreed that it would be meaningful to investigate these barriers and work towards obliterating them, paving the way for an expansion of the knowledge base. This points to the financial or economic viewpoint that has been prevalent in viewing the conduct of business entities, which is dangerously close to being biased and blind to other aspects of organizational performance. While organizational theorists have contributed to elaborate on why we should not be following a purely economic rationale in defining the sustainability index of firms, unfortunately, the accounting sciences are yet to catch up and innovate something different yet meaningful to bridge the gap. Accordingly, environmental concerns have remained the missing element of accounting expression.

ROLE OF CORPORATE ACCOUNTING TOWARDS ENVIRONMENTAL CARE

As the mechanism to translate how firms might interact with environment and generate different impacts is not covered within the accounting standards and relates, to some extent, to the overlooked responsibility of firms to deal with waste once it is pushed outside the organizational boundaries (sans legal compulsions), societal acceptance of such business practices and not being cognizant of the externalized liabilities, translate these impacts to remain outside the boundaries of accounting artefacts and an accepted reality. In this section, core accounting constructs and their failure at capturing the environmental implications of doing business are discussed, whereas further deliberations have been taken up in the respective chapters. Key points from those discussions are being covered here.

Financial Accounting and Reporting

The framework of financial accounting and reporting have followed GAAPs in developing the accounting interpretations of business transactions, which have mainly been sourced from the international accounting standards of the IAS, and in some cases the FASB. Here, a normative view of accounting theory defines the role of accounting records as capturing business events within the context of a socio-economic background based on the monetary considerations of business exchange. As waste and discards have traditionally been considered as valueless, these did not form a part of the inventory records or financial accounts. This is irrespective of any actions for the disposition of these assets. Why I am calling these organizational 'assets', contrary to their being considered as liabilities, is discussed in the sections 'Ownership of wastes and the tragedy of the commons' and 'New developments incoherent, yet a silver lining' and can be referred to separately. General accounting practices have transferred the material value of discards on to the finished products and services and to absorb material waste as a net change in the inventory at period close, successfully pushing these out of the organizational conscience. This also leads to the practice of recovering the loss of value due to waste by charging finished products and absorbing that directly, which otherwise impacts profitability but without pointing to efficiency, at least to begin with.

One of the arguments advanced for not considering externalities in accounting artefacts is the loss of causality linking organizations to the externalities generated by disposed waste, which is vital to establish or uphold in contractual obligations beyond organizational boundaries, save for the tort claims or contingencies arising out of environmental accidents. In the case of tort claims or contingencies arising out of environment accidents, organizational obligations are ascertained based on the potential damage that the firm is held accountable for. Other than that, social and environmental costs remain outside the accounting boundaries, within which the management, decision makers and stakeholders are offered information, obscuring the true potential, worth and risks of an organization. For example, the extended liabilities of an organization do not appear as a part of the balance sheet but still form a part of the financial risks that might rapidly expand within a dynamic environment, with any sudden change in statutes, practices and regulations. Secondly, the narrow view of accounting and its dependency on the contracting limits forces the management to act based on known considerations. In a bid to outperform markets and competitors, these decisions could introduce risky and environmentally unsustainable business practices aiming short-term gain, helping firms to appear as unconcerned corporate citizens to stakeholders and society.

With the advent of the IFRS as the global GAAP, there has been some advancement in the literature in examining how the principles-based accounting guidelines of the IFRS might help firms handle changing expectations of societies. Even though the research in this area is widening with the adoption of the IFRS by different countries, there have been limited efforts at examining the role of the IFRS in meeting the demands for sustainability or to change accounting practices. Accordingly, this sub-section looks to emphasize how moving from a country-specific GAAP to the IFRS might not change the primary focus of financial accounting and reporting. This includes examining the use of IAS 38 (impairment of emissions rights), IAS 32 (financial instruments: presentation), IFRS 7 (financial instruments: disclosures) and IAS 39 (the new IFRS 9) which collectively can be applied to deal with presentation, disclosure, recognition and measurement of financial instruments from carbon trading, some scholars have argued in favour of. Similarly, IFRIC 1 and IFRIC 5 from the former International Financial Reporting Interpretations Committee (now the IFRS Interpretations Committee) have been proposed to deal with the financial implications of setting up environmental funds that are related to decommissioning, rehabilitation and restoration events in case of certain industries in certain countries, whereas IAS 37 can deal with the provisions and contingencies associated with these events (Negash 2012). Similar assertions are made by Firoz and Ansari (2010), explaining how the aforesaid provisions of the IFRS could help firms report material aspects of SEA—for example, environmental reporting can reflect environmental assets (environmental rights and associated values purchased or contracted), actuarial value of insurance or similar risks, and capitalized values in environment-related research, while liabilities could relate to rehabilitation and restoration of impacted areas where physical data can be used as quantified aspects.

In-depth reading and analyses of these standards and analysing their implications reveals that (a) legal support is needed to reflect how firms might have chosen to deal with certain environmental issues (e.g., environmental restoration funds and compliance-related regulations), (b) market-based initiatives (e.g., carbon trading and emission rights) would need to carry a necessary regulatory infrastructure but still remain limited as to the implications that are of a financial nature and (c) opposing views put forth by scholars regarding the inadequacy of these provisions to handle different environmental assets and liabilities. The last point relates to, for example, how financial accounting would handle the initial carbon emission allowances and associated risks of recognizing these as a (non-fungible) asset, like IAS 38 has advocated. However, IAS 38 deals with non-tangible assets whose future economic benefits are expected to flow to the firms, but which is an issue to handle accounting of emission rights due to (a) an initial assignment not necessarily being a market-based allocation but the outcome of a regulatory process and (b) the non-fungible nature of emission rights not allowed to be adjusted against actual Kyoto (emission) units (Bebbington and Larrinaga-González 2008; Machado, Lima and Filho 2011). In continuation, IAS 37 is clear in its assessment that a provision (for liability) should be recognized by the firms only if they can assess a legal or constructive obligation because of a past event and if there is a probability of appropriating economic benefits generated due to the outflow of resources that are required to settle the obligation. This could be relevant in cases where the actual Kyoto units generated by a firm exceed the delivered allowances under the European Union (EU) carbon trading scheme (Sarkar 2010). This leaves little room to address externalities that are outside the market mechanism or in regions where the legal infrastructure is inadequate—for example, in developing and underdeveloped nations. In that sense, the IFRS' market-oriented approach to assess environmental performance due to carbon trading and other environmental regulations or obligations is hardly at more than arm's length from financial reporting under the local GAAP.

Cost accounting

Cost accounting is considered a bookkeeping mechanism of transactions that are internal to the organization. Accordingly, the IFRS or country-specific GAAP is generally not applicable for cost accounting, but still followed so as not to generate any unwarranted spikes in inventory valuation or cost of goods sold. As a matter of practice, accounting of internal transactions mimics the applicable GAAP to ensure consistency and maintain parity with the financial records.

Environmental Considerations and Conventional Accounting Theories

In advanced countries, compulsory maintenance of cost books of accounts has been done away with, although developing countries in places use regulatory measures to force businesses to maintain cost books of accounts. Where this is so, some form of accounting framework or generic guidance notes or principles are provided for firms to adhere and refer to. We can safely assume that the cost ledgers and information being produced therefrom cannot deviate significantly from the underlying GAAP. However, as the information produced from these books is for the internal use of the organization, there could be some experimentation to introduce environmental care as one element, or at least we can hope so!

Traditionally, while accounting for materials or in stock accounting, inventory accounting captures the material waste generated during the conversion process. However, as waste is valued at 'cost or realizable value, whichever is less', it would remain recorded in quantitative terms, without any significant value attached to it. The cost of waste generated would generally be loaded on to the outputs in cost terms, as we discussed previously. Once material waste is disposed of, though, quantity is updated in the inventory. Cost of disposal is directly charged to the financial books as an expense whereas earnings, if any, would form a part of miscellaneous income. Moreover, waste water and its subsequent treatment by the ETP plant might not form a part of the inventory books, unless stored separately and needing to be accounted for due to legal reasons. Emissions do not form a part of the inventory or cost books at all. This is to the extent of how aspects are treated. Next, if an organization deliberately wants to account for these aspects, would it be able to do so? Yes, it could. But the challenging aspect of it would be the rules that could be used to record these. Another area of concern would be the valuation norms to be used, and ascertain if the new valuation norms to be used would be any different from cost or realizable value whichever is least. Costing methods are traditionally designed to work with the definiteness of an underlying recipe, process or ingredients, which is an issue in cases where cost ascertainment needs to cross over the boundary of ownership or to include fuzziness or uncertainties. The latest costing techniques, such as life-cycle costing (LCC), full-cost accounting (FCA) and the total cost approach (TCA), have been advanced to overcome these limitations. Due to the advanced nature of these methodologies, only LCC is covered here. The other methods have been covered later in the text.

Life-cycle costing (LCC)

Life-cycle studies using life cycle analysis (LCA) or LCC methods are not new. These methodologies have been in use for quite some time. While LCC was designed by the US Department of Defense during the energy crises in the 1970s, these have mostly been used in the construction and defence industries (Korpi and Ala-Risku 2008). Scholars from this field are improvising upon these techniques to enhance their usage in environmentally sensitive decision-making and to promote widespread use of these methodologies (Gluch and Baumann 2004). Sterner (2001) has improved LCC to incorporate the considerations for green procurement in the construction industry, while Steen (2005) offered the perspectives of combining LCA and LCC and test the feasibility of developing a model that might enhance identification of environmental costs and benefits.

Like LCC, the LCA methodology was also developed in the 1970s in the US and deals in energy and mass balance computations, although not routinely used in organizations. It is easy to understand, it is not so popular within the accounting fraternity due to its characteristic of handling only physical data. However, LCA enthusiasts believe that it has substantial potential to capture environmental impacts and could support other methodologies as well (Steen 2005; Tukker 2000). Norris (2001) proposed integrating LCA techniques with LCC, resulting in an integrated tool to handle the life-cycle and economic implications of time-factored costs. By incorporating the economic implications of decision-making with LCA, it could improve its usability. Designed as full LCC, it could improve the on a traditional LCA by adding cost factors. Software tools such as PTLaser and TCAce are positioned to improve decision-making capabilities by taking environmental impacts into consideration. Traditionally, the life-cycle cost is derived as the total costs of an activity or object through the different phases of its life and computed in terms of its present value:

Environmental Considerations and Conventional Accounting Theories

$$LCC = \sum PV_t - PV_{\text{salvage value}} \qquad (6.1)$$

where, $\sum PV_t = PV_0$ ... Costs at the inception stage
$+ PV_{\text{project stage}}$ ... Costs at the project stage
$+ PV_{\text{maintenenace stage}}$ ... Costs at the maintenance stage
$+ ...$ ... Costs at other stages
$+ PV_{\text{end-of-life}}$ ... Costs of disposal at the end-of-life stage

Decisions depending on life-cycle studies would gravitate towards building a net-positive inflow for available alternatives, converting futuristic costs to the net present value (NPV) by using a suitably discounted cash-flow rate to opt for the one offering the highest margin.

Management (managerial) accounting

Traditional managerial accounting techniques involve comparing the costs with the benefits of a decision object, where all costs associated with the decision problem are compared against the benefits in the same or equivalent, generally monetary, terms. This poses a unique issue for the decision-making apparatus if it is expected to handle environmental care. First, can environmental care be expressed in definitive terms as a part of a decision-making problem? For example, can we define the environmental care that needs to be considered at the time of buying certain equipment or setting up a production unit? In terms of an investment decision, developing an acceptable level of environmental care has been a challenge. So the easier way would be to follow a legal compliance route and ascertain what a firm needs to adhere to, which could be the law of the land or may include market barriers that the competition might be faced with. In terms of financing decisions, what kind of environmental capture would connect or translate the risks associated with a decision? Strategic decision-making is an area where management accountants are involved to guide the management's short- and long-term decision-making with tools and techniques that emphasize a scientific approach to the problem and use information available within the organizational system. But what if the relevant information is not available with the organization or there is an inherent bias in the approach of the management accountant to work with certain tools and methodologies? Another concern in this area is whether the organization is ready to factor in environmental questions as a part of its decision-making process. Let's examine this through a sample case:

> Example: ABC Inc. is planning to install an industrial boiler of capacity 5,000–10,000 kg hr^{-1}, with other specifications such as pressure, make and model. All supplier quotations are being evaluated for technical as well as commercial details. The corporate controller, who is a management accountant by profession, is helping with the commercial evaluation of proposals while the chief of maintenance and a plant engineer are tasked with jointly evaluating the technical aspects of the proposal. Traditionally, capital investment-related proposals are evaluated in the organization using the return-on-investment (ROI) or total cost of ownership (TCO) methods. Accordingly, the evaluation process involves minimizing the investment and maintenance costs.
>
> Problem: How will the evaluators consider the environmental overheads (to start with) and how well can the underlying methodologies incorporate greening details, such as the environmental compliance ratings of suppliers, the environmental performance of the machines and related trial-run data, and energy and green product ratings of the equipment?
>
> To reduce the per unit cost of steam, a firm might work towards replacing its coal-fired boiler with a biomass boiler. So the theoretical cost of operations would consider the accounted data but not the environmental impacts that it generated. Accordingly, the decision could primarily be governed by the economic priorities and *not* improve environmental impacts. The resulting environmental gains, if any, would only be the by-product of the economic decision. In other words, if the cost of steam remained the same in both cases, would the firm replace its coal-fired boilers to reduce the environmental impacts?

Traditional theories on decision-making are rooted in the concept of better performance (in economic terms) being a fiduciary duty of the management, along with increasing the market worth of the organization (per agency theory). Moreover, traditional management accounting techniques are not designed to evolve and accommodate decision variables that are not or cannot be expressed in economic or financial terms. Accordingly, management accountants would be compelled to force-fit all other decision variables to a single paradigm that supports a predominantly economic viewpoint.

Organizational externalities and limitations of traditional accounting practice

Traditional accounting techniques are dependent on the generation and interpretation of information on transactional value. While cost accounting refers to the methods and techniques to capture transactions and ledgerize costs by following an internal value-creation process, management accounting uses methods and techniques to generate information and help users interpret these within a valid decision-making context. The traditional role of waste analysis in costing and estimation techniques is limited to planning for the procurement of materials and services and driving process efficiency by controlling waste within the defined boundaries. Similarly, waste resources (say, idle hours) are used to calculate standard loading profiles of resource costs on to the products or processes and help the firm earn overheads as a part of the performance (Horngren 2004). Accordingly, even though the costing system collects information on waste materials and resources, its usefulness is limited to driving process efficiencies.

$$\text{Total input loss (\%)} = (\text{Input quantities} - \text{Eq. output quantities}) \times 100/\text{input quantities} \qquad (6.2)$$

Note: Eq. output means equivalent output.

As externalities take place outside the accounting books and beyond the organizational boundaries, traditional accounting frameworks would be constrained to generate any information on these. In other words, accounting constructs can hardly generate information different from what they have been designed for. However, the changes needed within the accounting functions to handle the environmental challenges of an organization would need to be fundamental to how organizations use accounting, a topic to be debated next. Since environmental concerns are not captured in the traditional forms of accounting and costing, it would be premature to imagine the use of conventional accounting and decision-making methodologies can involve environmental perspectives.

WHERE DOES THIS LEAVE US?

Traditional accounting practices offer methodologies to record transactions, analyse costs and develop an information base for the management to improve the profitability of the firms. By comparing the temporality of captured information, practitioners help the management take informed decisions on improving the utilization of materials, funds and resources. However, the way accounting captures events are enacted through a mathematical construct that is grounded to narrate the story of businesses in time by using monetary values only. So even when firms behave as responsible corporate citizens and are serious about improving their social and environmental contributions, the question worth asking is whether accounting can support them any better than how it is doing today. This question is crucial in tracing the advances in accounting theories in the last two decades, to answer questions like why developments followed a specific path against other possibilities.

Needless to say, the normative theory of accounting allows us to view accounting as a framework that encapsulates the fundamentals of organizational mechanisms within certain guidelines, procedures and standards to aid practitioners (accountants) in converting the social experiences of firms into a temporal repository of events that becomes a primary source of information in tracing the economic history of the firm. Being mathematical in nature, the construct uses the language of bookkeeping to interpret business transactions by abstracting the social experiences of firms into a two-dimensional space of time and value (historical costs) that does not have a place to capture any other variable, including the environmental or social implications that the critical theorists have argued for. Accordingly, even if for arguments' sake, a firm is ready to analyse its environmental fingerprint, the language of bookkeeping and the normative aspects of accounting were not developed to translate the impacts of business events readily into available information, nor has it been found practical to extend the existing construct to accommodate the new information needs of calibrating environmental and social impacts. Since this limitation is perennial, the question is whether there a way to circumvent it. Unfortunately, the answer cannot be a simple 'yes' or 'no'. This is one of the core areas of investigation followed through in the ensuing chapters.

Contemporary Developments in Green(ing) Accounting

INTRODUCTION

Environmentally conscious accounting has been a subject of intense debate in the last two decades, enriched by the contributions of scholars and practitioners from around the world. However, being an emerging field of scholarship, it still has to address concerns like multitudes of definitions, overlapping ideas and lack of a taxonomy. Environmental (or green) accounting is the nomenclature used to reflect the advances in the system of national accounts (SNA) at national/regional level as well, to be able to integrate environmental considerations of the region into its macroeconomic data pool. Within the parlance of the SNA, green accounting is expected to reflect spatio-temporal insights on growth strategies pursued by the region and their impacts on regional resources. On the other hand, at the micro or the firm level, environmental accounting has been a part of the theories and models to study and reflect upon the behaviour and responsibilities of firms, and how environmental considerations are addressed by them. Studying these two parallel lines of scholarship helps us to reflect on the overlaps among the theories and how ideas from the macro and the micro domains are cross-fertilizing each other. However, as yet, advancements in the micro and the macro world do not have a middle ground to converge upon. In other words, while ecological disturbances and the corresponding reasons are being connected to anthropocentricity, connecting these to micro- or unit-level behaviour and business activities remains. Given these concerns, this chapter reflects on emerging trends in this exciting area of scholarship and the new theories that are forming under the aegis of environmentally sensitive accounting.

ENVIRONMENTAL (GREEN) ACCOUNTING: THEORY IN THE MAKING

Even though the field of green accounting at the regional or macro level is a relatively new area of development, insights from the literature are included here to help the reader appreciate the complexity of problems and to build arguments on how firm-level activities can be aggregated to study the influence they have on the region and to develop parity between the local and regional areas of growth by leveraging ideas across this divide. For example, rapid industrial development and high growth rate of industries have been fruitful and contributed to the gross domestic product (GDP), but have also resulted in high level of pollution, a case in point being the haze, smoke, smog or high concentration of particulate matter (especially $PM_{2.5}$, PM_5 and PM_{10}) that the industrial towns and developing worlds are now living with, for example, given the perennial unhealthy quality of air and smog[1] in Shanghai, Beijing, the industrial belt of North India and New Delhi. Unfortunately, these challenges have failed to find reflections in the development goals or growth models of the respective regions, constantly being overlooked by the global south in a bid to 'quickly' reach the affluence levels of the global north, guided by the economic theories that do not have a place to accommodate social and environmental realities or to factor in deterioration due to industrial activities. At the same time, developmental agendas are leaving their mark all over the place, with the loss of natural habitats and forest cover, the extinction of species that now have to compete with a growing human population for space and natural habitats, and shrinking space for wilderness, being the losses left outside the growth theories.

Green accounting at the regional or macro level

Scientific studies of the Anthropocene have developed the knowledge base of human activities, the built environment and their interactions

---

[1] *Under the Dome*, a self-financed Chinese documentary (narrated by Chai Jing, produced by Ming Fan) offers an excellent visual narrative of this challenge, including how doctors are finding it difficult to create lab conditions to expose controlled groups to breathe in 'healthy' or 'normal' air and study them to establish scientific evidence on respiratory conditions under continuous exposure to unhealthy air (Halton 2015).

within and across different subsystems. Applying the principles of the ecological sciences and of economics, environmental economics evolved models to capture the stocks and flow of materials, energies and natural resources (with some level of granularity) within a region, and to cross-link these with economic activities within the region to profile the spatio-temporal impacts on the natural habitat. Developments include engineering challenges that have aided in technological improvement and expanded the growth and progress of human societies. In the context of the industrial ecosystem, including symbiotic relationships and feedback loops, 'industrial ecology' is being developed as a subject that relies on the systems approach to examine industrial enterprises and their interconnectivity as it relates to social metabolism, rate of production and consumption, retainage and disposal. Although deeply rooted in the general systems theory and aptly supported by computing power to process data and map industrial flows, these studies are limited to human activities and their interactions with nature at different points when viewed from a closed-system perspective, instead of relating to how an open system might interact with natural entities (Gradel and Allenby 2003).

The accounting of natural resources at an aggregate (regional or national) level is parallel to the traditional framework of the SNA, which aggregates goods and services transacted at the regional or national level and can use established monetary equivalency to reflect the anthropocentric sense of value that we are used to (Rout 2010). Within the SNA, non-market driven ecosystem services such as carbon sequestration, flood controls, and measures taken against deteriorating air quality and soil erosion are generally ignored, in favour of tangible ones (or use value) that are derived from the natural resources. Accordingly, environmental degradation is not reflected within the SNA. The traditional flaws of the SNA can be countered by using different approaches such as pollution expenditure accounting, physical accounting, development of green indicators and extension of SNA-type systems such as the system of environmental-economic accounting (SEEA). These advances range from improving the existing constructs to record and generate new information to stand-alone constructs that run parallel to and independent of the existing ones. Pollution expenditure accounting is the restatement of already accounted information on the GDP whereas physical accounting involves accounting of the physical stock of natural resources that has undergone a volume change—for example, the system developed in the Netherlands as the National Accounting Matrix including Environmental Accounts (NAMEA). Physical accounting of resources needs to deal with multiple accounts and has large data sets to record. An alternative is a green indicators approach, which can be developed in line with social indicators to measure social well-being in addition to industrial progress.

Vannella (2004) has studied the difficulties in developing integrated systems for economic and environmental accounting at the regional and state levels. For sustainable development of a region, it is mandatory to develop environmental variables that would carry the impact of the human population within the region. SEEA separates and elaborates all environment-related flows within an economy to develop the physical accounts of its natural resources. Extensions of an SNA-type system can create satellite accounts, which could be another way forward to developing environmental or green accounting. SEEA and the Environment and Natural Resources Accounting Framework (ENRAP, also known as the Perkins framework) are two methods that are in use in environmentally advanced countries (Rout 2010). As compared to frameworks based on extending SNA, if the NAMEA project is used as a platform to identify the costs associated with monitoring, controlling and repairing environmental damage at the regional level, this would need the development of a registry of variables through which the state of health of the environment could be measured. Alternatively, a coexistence model could be used—for example, a study of environmentally polluted areas along with their economic well-being—but this would need the development of a database to monitor the environmental plan, the spatio-temporal profiles of the environmental aspects and presence of strong national bodies to participate in the research and draw inferences. However, as complete knowledge of interactions within the entire ecosystem is not available, it would need progressive work to develop meaningful observations and draw future developmental plans (Vannella 2004).

Pedersen and de Haan (2006) have improved the interlinking of the SNA and the SEEA by developing physical input–output tables (PIOT) to demonstrate the economic relevance of material flow accounting (MFA) within the economy. The SEEA has developed environmental functions for an economy to extend the SNA by using additional accounting constructs. It also includes accounting for environmental protection activities, valuation techniques for measuring environmental depletion of natural resources, and relate it to the SNA. However, as the goal of balanced development is to study the industrial and economic growth with the consequential loss of natural resources, it has been found to be difficult to connect the two without developing multidimensional non-linear causal relationships, whereas PIOT connects the physical quantities of input–output for multiple industries and generates data for analytical purposes by using interconnected equivalencies. Another way to handle the shortcoming of connecting economic and industrial growth with loss of natural resources would be to let the SEEA's physical account approximate to MFA-type accounts and add the indirect flow of materials as an accounting extension (Pedersen and de Haan 2006). Pedersen and de Haan have created a complete PIOT for an economy, based on the earlier work done at a few of the EU countries such as Denmark, Germany, Finland, Italy and the Netherlands, and integrated the SEEA with the SNA. Their 'full PIOT' framework refers to the full description of physical flows and use of specific terminologies to indicate the type of flows being addressed. On the other hand, developing countries such as Indonesia, Costa Rica, China, the Philippines and Brazil are experimenting with more a conservative version of the same theme, using a model promoted by Repetto et al. (1989) where, based on the mass and energy relationships, the environmental impacts can be linked to specific types of economic activities.

From these discussions, it can be inferred that green accounting at the regional level contributes to the debate on environmental protection versus economic growth, relates the environmental impacts to the policies being followed and opens up links between environmental and economic policies. The research in this area is shaping up as a separate line of study within ecological economics, with shared terms and references which could be leveraged at a firm or entity level. Yet, as noted earlier, environmental challenges and related developments at the micro and macro levels still do not find a common ground, nor exchange developments across domains. A deeper look shows that the traditional framework of accounting at the national level is facing the same set of concerns as firm-level accounting, revealing the deep roots of economic thinking as a part of human activity in general.

Environmental accounting at the firm level

Similar to the advent of green accounting at the regional levels, environmentally conscious accounting at the corporate level is not a well-defined subject either. Once the question of unsustainable practices in the industrial world was established, researchers and thinkers questioned the economic viewpoint of accounting and its role. The concept evolved as a set of changes expected within the accounting functions of a business enterprise to develop green thinking in a firm. However, most of these advances have happened in silos. The previous chapter has discussed the ineffectiveness of accounting due to the boundary limiting its considerations and the need to study the feasibility of connecting accounting to environmental sustainability. Still, the growth of literature can discern these ideas into distinct line of thoughts, outlined below:

- The first one is the use of organizational disclosures to disseminate policies, programmes and practices being adopted by the firms towards the greening of products, processes and services across different channels, such as annual accounts, webcasts, investors' forums and corporate announcements. Several articles from different theoretical schools have studied firms' extents of disclosures and linked it with different performance criteria such as financial performance, market response, organizational intent and others, to draw up empirical rules governing the behaviour of firms.
- Another line of thought has evolved around exploring the already accounted data to help management with information on waste and the associated costs as a part of EMA. This relates to the methodological solutions with improvised accounting techniques and explored further in the section 'Environmental management accounting'.

- A third school of thought has emerged in the form of SEA and offers a philosophical basis for developing environmentally conscious accounting in theory and practice.

These developments are expected to address the changes needed in the accounting artefacts, flow of information, decision-making paradigms and the attitude of accountants. However, the developments within the accounting sphere are yet to be assimilated into theories to neatly articulate environmentally conscious and socially relevant accounting practices. The remainder of this chapter reflects on these advances.

SOCIAL AND ENVIRONMENTAL ACCOUNTING (SEA)

SEA has evolved as a study with advances in the accounting theories to include social and environmental considerations as a part of conducting business. In the last 20 years, a substantial body of work has evolved under this umbrella concept and covers a considerable breadth of discussions on different aspects of the subject. At a philosophical level, the concept of financial reporting and its economic implications can demonstrate the implicit assumption of firms to move from a less economically desirable state to a more desirable one. Basic accounting functions convert the experiences of social construction through a measurement scheme that reflects such desires, but that has not been a successful way to promote environmental and social welfare, nor the accounting process has any construct or capability to reflect it. It could be argued though that unless accounting is able to provide social justice for both, present and future generations—that is, promote equitability—it would be meaningless to assume it will contribute to sustainable values. Still, there are various views to explore in this area:

- Based on the rights of society over the allocation of resources and the responsibility of economic entities to keep stakeholders informed, SEA challenges the conventional wisdom and considers accountability to be a part of accounting convention (Gray 2008). This is a reflection of the *ethical rights theory* discussed earlier.
- Brown and Fraser (2006) have provided an analysis comparing three approaches in SEA—the business case, stakeholder

accountability and the critical theory approach. The business-case approach covers the corporate social responsibilities (CSR) that view SEA as a means to develop a win–win relationship between businesses and stakeholders (or the *instrumental view*). In this case, CSR and SEA are extensions of the managements' toolkits to enhance shareholders' wealth and avoid a conflict of interests in the process.

- The stakeholder-accountability approach is based on the theory that large organizations are quasi-public and responsible for promoting an open, transparent and democratic society (the *legitimacy view*). This view brings in the accountability of businesses towards developing social and ethical risks, which helps the firms to demonstrate their commitment towards their responsibilities. While the business-case approach seems to suffer from picking low-hanging fruits, the stakeholder-accountability approach suffers from organizational concerns about possible image makeover.
- The critical-theory approach, in contrast, is radical and refers to the worrisome conduct of capital orientation within the business environment, which force the organizations and SEA policies to fall prey to these tendencies (the *ethical approach*).
- Taking the *methodological view*, Bebbington, Brown and Frame (2007) believed that SEA needed to explore the means to cover monetary and non-monetary aspects of the social and environmental impacts of businesses for it to improve its dialogue with the stakeholders. Although monetization was not always a fully developed option, then or now, authors believed that it needed to improve the logical form of accounting to enable collection of information and its dissemination for the debate and views of a polyvocal citizenry.
- Fraser (2012) believes that the SEA and its technologies are in a better position to influence the internal operations of organizations and contribute towards sustainability-related accounting applications. SEA technologies could also play the role of a catalyst to change the behaviour of a firm and be a part of an assemblage to continue the process of change process that might be reactionary (morphostatic) in nature. Eventually, the expectation is that these changes could lead to morphogenetic or second-order changes,

where changes leading to long-term impacts are considered as more important and bring sustainability-related issues to the forefront of organizational existence (the *instrumental view*).
- On a voluntary basis, companies would prefer to use SEA for internal reporting rather than external reporting, as gathered from a survey of Australian companies (Herbohn 2005). The surveyed firms modified the existing accounting system to improve environmental and social management accounting practices. The companies with the best practices could use cost accounting data of environmental and social impacts to evolve internal decision-making on cost savings and comply with the internal and external reporting requirements (such as that of the Global Reporting Initiative, or GRI). However, the practitioners believed that the lack of general theorizing within the SEA theories evidence the need for further work that might require engaging practitioners' perspectives to bring about changes within the accounting practice (Parker 2005) (*instrumental view*).
- Philosophically, SEA can be viewed as a platform or point of view that establishes a dialogue between the society and the organizations, in order to help firms and their management to detail how they believe firms can be sustainable. Even though SEA can only be insufficiently represented through financial units alone (to reflect methodological levels), the need to disclose relevant information in financial (monetary), non-financial (monetary and physical units) and/or mixed units will help firms to develop an honest and open dialogue with society, which would enable the firms to move from 'trust me' (a disclosure-based approach) to 'show me' (achievements). The effectiveness of SEA can also be defined as the extent to which it meets the objectives of maximizing social and organizational benefits and meeting the financial and non-financial information requirements of internal and external stakeholders of businesses (Orlitzky and Whelan 2007) (combining the *ethical approach* and the *ethical rights view*).
- One of the experiments within SEA was to develop unified social and environmental management accounting system and to improve social and environmental accounting reporting (SEAR) that works in coordination with the existing management accounting system.

In the absence of discussions on social and environmental information, biases could be introduced in the decision-making processes that might lead to information asymmetry and risks (Nikolaou and Evangelinos 2010). To develop such a system, a conceptual model was proposed by Petcharat and Mula (2012) where overlapping and interlocking regions of economic, environmental and social accounting perspectives could become a basis to develop a pragmatic approach (*instrumental view*).

- Although there has been less success in generating financial equivalents to SEA, it might increase the relevance of socially responsible decisions, it was felt, which would need to involve accountability towards sustainability, a substantial change in the society–firm relationship. This can be improved by using SEA as a medium to report risks and uncertainties associated with the business. Lack of information about risks is one of the main weaknesses of the current accounting literature. Even with the disclosure rules, empirical studies have found large variations and deficits in risk reporting. This can be related to the uncertainty of information endowment and lack of credible communication. The risk-regulation framework encourages companies to disclose financial risks, but other than the market risks, all financial and non-financial risks rely purely on voluntary information (Lungu et al. 2009/10) (*instrumental view*).

From the details above, we can reflect how SEA theories have gone all over the place as they attempted to fill in the theoretical foundations for a solution (instrumental view) that can connect accounting with stakeholders' needs. It also tried to become a medium to reflect the environmental sensitivity of firms, to contribute to the dialogue with society and stakeholders. This envisages the role of SEA as an enabler (artefact) as well as the enabling process (medium). Scholars have also viewed SEA as a tool for short- as well as long-term changes in the behaviour of firms. This view is closer to Gray and Laughlin (2012), who discussed the development of SEA as an independent branch of research by challenging the underlying accountability of existing business practices. Besides discussing disclosure and reporting practices, they also highlighted the tensions within the 'light green' and 'deep green' environmentalism and between regional managerialism and incrementalism versus the deep and fundamental change needed in the methods of economic organizations. Parker (2011) reflected on the subject of the advancement of research to balance 'environmental' versus 'social' subjects and the general trends in developed economies who have embraced environmental considerations at the national level (such as Australia, Finland, the Netherlands, New Zealand, Spain, and the UK). However (from a regulatory perspective), these developments could have been for reasons other than the respective governments subscribing to SEA theories. As for compliance, I do not believe that outcomes were always driven by SEA theories, although SEA theories seemed to embrace the developments ex-post. Other than disclosure requirements which the theorists believed would enhance accountability, SEA did not offer any practical guide to improving accountability, neither did it theorize how accountability can be improved through voluntary disclosure practices.

ENVIRONMENTAL ACCOUNTING— DISCLOSURE PRACTICES

Disclosure is the process of disclosing facts or information that have been withheld earlier. The practice of disclosing environmental impacts that a firm might be producing or the managements' plan of action to counter these impacts can be seen as being in favour of accepting societal rights to know including how the scarce resources are being used—this has its roots in the stakeholders' theory. Political economy theories view disclosure practices as firms' response to pressure exerted by different stakeholders' groups, whereas stakeholders' theory defines these as a motivation for strategic and ethical responsibilities, where disclosed information can be interpreted (in some cases) for its usefulness in decision-making, economic performance and contribution to social and political theories. Nevertheless, this upholds the view that firms are quasi-public institutions in nature and act as conduits for society and its inhabitants to achieve social aspirations. At the same time, ethical views challenge older understandings of the purpose for which firms exist, so that now firms and their activities can be considered open to public scrutiny, where the activities and outcomes are in the public space. This challenges the view held earlier of firms being a closed system where the boundary of interactions between firms and society becomes less deterministic to isolate the firms from the socio-politico-environmental space within which they are operating.

The literature favoured the stand that firms need to legitimise their businesses by informing stakeholders about how they are acting or reacting towards environmental perils, upholding the stakeholders' right to know. The legitimacy theory uses disclosure practices as strategies for firms to seek acceptance and approval for their activities from society. Moreover, the diversity of results from these studies hardly generated any hard evidence to investigate the impacts of corporate communication practices. Invoking the ethical rights theory to understand the development points to the firms' compliance behaviours to societal demands, mostly the ethical positives. To be noted, ethical positives are the information on the actions and steps taken by the firms to stand on the right side of the ethical divide—that is, what the firm has chosen to do, how it is planning to do so and why the firm thinks this is the correct course of action. Again, if the discourses are not legally contoured and instead it is left to the firms' discretion what content it chooses to disseminate, more often than not, only the ethical positives would make the rounds.

Research into environmental disclosure practices in different countries[2] has not found any uniformity in patterns.[3] Although the surveyed literature is not devoted to covering advances of environmental accounting within the firms, they reflected a shared taxonomy and viewed the organizations' disclosures as a sign of the recognition of their responsibilities by the firms and of taking cognizance of their

[2] For example, in Australia (Gadenne and Zaman 2002), Europe (Moneva and Ortas 2010), Fiji (Lodhia 2003), India (Malarvizhi and Yadav, 2008/2009; Singh and Joshi 2009), the Philippines (Rao et al. 2009), Spain (Llena, Moneva and Hernandez 2007), Sweden (Nyquist 2003), Taiwan (Huang and Kung 2010), Thailand (Robkob and Ussahawanitchakit 2009) and the USA (Dorweiler and Yakhou 2005; King and Lenox 2001).

[3] Google Scholar lists 3,500 articles with the keyword 'corporate disclosure' in a title search since 2000, which jumps to 20,000-plus articles if 'corporate' is dropped from the keyword list, giving an insight into its popularity in accounting research.

behaviour towards society. For example, Spain showed an improved quantum of environmental information in the annual accounts of Spanish companies once environmental disclosures were made mandatory by the Instituto de Contabilidad y Auditoría de Cuentas (ICAC, the accounting and auditing institute of Spain). The findings were achieved by comparing the annual reports of the firms ten years apart, that is, before and after the issuance of the directives (Llena, Moneva and Hernandez 2007). However, similar improvements were not observed within Indian organizations in two studies conducted three years apart and involving two different dissemination channels—internet-based reporting and annual accounts of the listed organizations—without any corresponding change in the mandatory accounting norms (Malarvizhi and Yadav 2008/2009; Singh and Joshi 2009). In a similar study from Thailand, voluntary disclosures made by organizations from the food and beverage sector were found to be correlated to multiple factors, such as corporate image, customer acceptance and the firms' commitment to sustainability (Robkob and Ussahawanitchakit 2009), whereas the study of environmental disclosures of 759 Taiwanese firms could not come up with any definite conclusion regarding the interrelation of financial performance and environmental disclosure (Huang and Kung 2010). Considerable attention has been paid to establishing, analysing and predicting a relationship between performance and disclosure practices in some of the studies (Eugénio, Lourenço and Morais 2010). However, Branco and Rodrigues (2007) have rightly pointed out regarding the methodological difficulties associated with the interpretation of empirical data from the disclosure-based research reports, including the inability to measure the extent of the disclosures or to endorse the quality of disclosure to reflect the advancements of the firms in addressing social and environmental externalities.

Another pertinent question worthy of attention is why accounting-based theories suddenly started endorsing one or another form of disclosure-based practice, including countless efforts to decipher what was being disseminated or how much of it could be related to the environmental commitments of firms. However, there is a little corroboration for an association between the use of voluntary disclosure and changes embraced by the firms, or its use in the decision-making areas. Looking back, research involving disclosure-based studies seemed like a wave in the scholarly pursuit to discover empirical validations between level of disclosures and environmental care so as to predict environmental sensitivity of firms based on the quantity and quality of disclosures, without realizing the impossibility of establishing such a correspondence, simply because most of these disclosures are ethical positives, driven-off of different agendas. Developing a balanced view would need the framework to call for ethical negatives as well and evolve a comparative approach that can be used to discover true nature of things.

This is where we draw insights from the ethical rights theory to explain that accounting and reporting today provides one-sided information, mostly the economic positives that firms find easy to disseminate following regulatory guidelines and/or otherwise in voluntary terms that limits the field of view of the users of the information, especially in absence of corresponding negatives that firms are not bound to produce, thereby contributing to information asymmetry. Unless the claims of positive action are backed by information on the negatives in every case—for example, quantified information on waste, emissions, ethical violations, labour unrest and the efforts of firms to counter these and to discharge their social and environmental obligations should be made available to substantiate the claims of a firm being socially just and environmentally sensitive—it might appear as a deliberate attempt on the part of the firm to promote a specific, perhaps hidden agenda. Extending from the stakeholders' theory, the stakeholders' rights to be informed about the negative impacts that the businesses are contributing to has not been exercised diligently by the stakeholders, resulting in letting the negative impacts remain as implicit to the ethical positives, causing more harm to the society. The argument here is that to be reflective of the real worth of a firm, both sets of information (ethical positives as well as negatives) are a necessity and it should be feasible to assess these independently from the current practices of the firm. This will help the stakeholders to evaluate how firms are contributing to the overall well-being of society and nature, not to mention a better discharge of its accountability. However, other than the organizational intent, this would also depend on methodological advances that provide methods, tools and techniques to help the cause. This is where EMA and related advances come in, and are explored next.

ENVIRONMENTAL MANAGEMENT ACCOUNTING (EMA)

EMA is the umbrella term used to refer to the theory and practice of institutionalizing changes within the accounting practices (internally) to enable firms to become environmentally conscious, by leveraging environmentally enhanced methods, tools and techniques and improve their accounting and decision-making capabilities. The primer of the USEPA (1997) was the first document to propose environmental cost accounting (ECA) as an extension of the corporate accounting framework for organizations to prioritize environmental considerations in accounting and information generation. Although accounting has claimed to have an armoury sufficient to handling environmentally sensitive costs, the GAAP did not cover costs beyond the organizational boundaries to address the larger concerns around the environmental impacts of different business activities. This in effect means that the formal accounting construct was not ready to change, to enable the accounting practitioners to interpret business events to decipher how they might be impacting the environment and left the onus on the internal arrangement of costs. This philosophy of not changing the formal accounting constructs is derived from the economics of business, which shields firms from interpreting business activities no differently than they have been practised in the past. For the time being, the governing bodies that deal with accounting standards might sideline these demands as these are not hurting the majority or the core of businesses, similar to how externalities have been viewed by economic theories; however, they might find themselves in a fix soon when the externalized liabilities have outgrown the contributed economic value. As a result, like with the earlier development of cost accounting that grew due to the need to relate the internal movement of costs to value creation, uncovering environmental impacts has somehow became a problem of rearranging internal costs. This short-sighted view is based on reducing waste and discards, those being the main source of impacts. Consequently, the logic goes that if waste and discards can be reduced, environmental impacts would also be reduced. Further, it is suggested that an effective way to reduce these would be find out the related expenditure class, which might need deeper investigation to control. All these efforts were brought within the umbrella term of 'EMA'. A review of the state-of-the-art literature on the topic has been covered in the next chapter. This chapter is more about analysing the major ideas covered through the development of EMA and to reflect on the purpose, which is to un-cover environmentally sensitive physical (energy, water and materials) and financial (covering environment related costs, earnings and savings) data and provide information to the decision makers including the stock and flow of materials, energies, wastes, water or any other environmentally sensitive element that is being used, consumed and/or wasted to manufacture and market goods and services (IFAC 2005).

Within the EMA framework, several methodologies have been proposed and experimented with by the researchers and practitioners. These include the waste accounting and reporting methodology developed under the aegis of the United Nations Division for Sustainable Development (UNDSD) to help firms improve their environmental performance by analysing overheads from the books of accounts using environmental cost drivers (UNDSD 2001). The development of accounting-based methods such as material flow cost accounting (MFCA) and waste accounting used accounting-based constructs but without venturing beyond the organizational boundary to define environmental cost (IFAC 2005; Kokubu and Nakajima 2004; MOE 2005). Other non-accounting decision-making tools such as life-cycle methods, FCA, environmental performance indicators (EPIs) and TCA have been experimented with to incorporate costs beyond the organizational boundaries and contributed to building alternative perspectives. Some of these methodologies were experimented with as part of EMA implementations and others in isolation, but the diversity in accounting techniques as a part of these experiments and the lack of taxonomy remained.

Jasch and Lavicka (2006) believed that as compared to the control of waste through internal costs, environmental externalities should be handled as part of a different construct such as environmental accounting, wherein different valuation methods (contingent valuation or the avoidance and restoration method) can be legitimized to help organizations monetize externalities. However, the authors believed this was not the responsibility of firms and that it should be left to the corresponding regions to develop interpretations that would work better with policy initiatives. Still, advances within EMA have led to the identification of new areas of cost measurement, development of improved methods of cost analysis, rethinking of strategic decision-making approaches and building of new performance measurement systems, for example, by using measurements such as eco-intensity (which has been covered in 'Methodological Developments in Environmental Management Accounting'). Pulver (2007), however, challenged the conventional wisdom that the developing countries are environmental laggards and are responsible for the poor environmental upkeep of business practices. It is equally true that transnational organizations have been shifting polluting industries to the developing nations but without transferring the technological know-how, which saves them production costs as well. As the environmental laws are yet to acquire stringency in economies of the global north, lower technological capabilities of the global south might contribute to the higher pollution levels. In addition, the closer collaboration of market- and export-led growth has led firms in these countries to adopt a shorter route, like certifications to validate the seriousness of their efforts. It would need multiple areas of cooperation such as knowledge sharing, collaborative research, and technological and political collaborations to bring environmental stewardship in the global south in parity with the global north and improve the discharge of environmental responsibilities as a part of business conduct (Rosendahl and Strand 2011).

ACCOUNTING IN ITS ENVIRONMENTAL *AVATAR*

To bring environmental considerations within the purview of accounting theory and practice, there have been multiple threads of development within the contemporary literature that overlap substantially. The developments have evolved parallel themes of research and experimentation, evidencing the overlapping use of terms, a multitude of definitions and intense debate on different concepts and practices, most of which have remained influenced in part by corresponding advancements within the economic theories. The environmental considerations of firms (if any) could be reflected through their inclusion within their disclosures as well as developing standalone constructs to record information on environmental impacts and costs, which would hardly be surprising, as these efforts are natural extensions of the feasible solution space.

As with any evolving area of study, these overlaps are yet to find convergence. Similarly, the efforts of greening the SNA are a work in progress and have remained disconnected from the micro-level efforts. Overall, the greening strategies at the micro and macro levels have remained independent developments, and efforts are needed to develop connectivity between the two realms. At the micro level, critical perspectives of accounting have shaped the views that the SEA theories are reflective of, for example, the expanding accountability of firms within the organizational conscience, hoping for an eventual breakthrough to develop an expression that would be inclusive within the accounting practices and lead to morphogenetic changes. Persisting with the need to follow an ethical approach is the highlight of the SEA theories, which believe in organizational commitments towards the environment and to society as being the right thing to do, as if firms are living things that have a moral fabric and can keep on learning from surrounding events and their participation in them and would eventually become accountable towards society and environment.

EMA has produced several methodological improvements to isolate waste and the costs associated with it. The techniques included cost computation methodologies that ranged from fixed-recipe computations to one that could handle uncertainties of information and assign costs to waste using economic arguments to contribute towards reducing the overall waste flow. However, if cost accounting and costing norms are 'regulatory-controlled' (as in case in India and other developing countries) they would remain insensitive to the demands of changing times and their inability of conventional costing techniques to respond to the evolving needs of the organizations is beyond argument. However, if the firms could generate a business case using environmental care as the primary driver, they could expect higher levels of social acceptance and customer loyalty, notwithstanding how the externalities are calculated. Qualitative solutions such as disclosure-based practices evoked the role of firms in improving organizational commitments towards external stakeholders.

A part of this trend emerged as voluntary reporting mechanisms that would use standard as well as non-standard formats to report waste and other environmental aspects within the umbrella term of CSR, which somehow salvages the situation without bringing in regulatory and legal routes. These efforts remained voluntary, as legality was believed to limit the efforts and discourage firms not to go beyond the regulatory norms, whereas a command-and-control approach was believed to be counterproductive for enabling firms to evolve consideration towards regional and social concerns as a process. Moreover, introducing legal measures would also means using taxpayers' dollars to set up effective governance and penal procedures, so it is believed to be a costlier proposition. For example, consider the case of India, where a legal statute has been put in place to define what CSR is or could be, by enacting CSR rules under the Companies Act 2013. The law is clear in defining what can be considered as a CSR activity (schedule VII lists specific types of activities) and the mandatory character of such an intervention should be (2% of average net profit during the three immediately preceding financial years), and it is a mandate for firms to be engaged in CSR (all public limited companies with a net worth of ₹5 billion, a turnover exceeding ₹10 billion or a net profit of ₹50 million). Here, instead of holding firms accountable for the quality of a value addition, a safe passage has been provided in the form of an economic programme where a specified percentage of their profit gets employed in a characteristic manner towards funding 'escape routes'.

Besides offloading the social responsibilities of the government to the businesses in the name of CSR, the mechanism can also be viewed as allowing firms to balance one wrong (ignoring social and environmental perils) with another right (unrelated social projects) that is entirely unrelated, leaving aside the basic question regarding the quality of the value addition to the environmental and social capital that the firms are enhancing or depleting. At the same time, this introduces 'economic thinking' as a part of public governance, which avoids fulfilling the promise of equitable distribution of natural resources to the masses by developing a raison d'être to circumvent it altogether. As these developments suffer from the shortcoming to achieve a possible enactment beyond models and frameworks, the onus has shifted to the fragile conscience of the firms (as if they have one) to decide how they might want to pursue the choices (as if they would support altruistic reasons), while hoping for them to behave in a way that would optimize the well-being of all concerned. In a case, the best practical outcome would be a win–win situation where environmental benefits are incurred through the economic rationale, although I am not sure how workable or practical that might be. Even though this might be a representation of opportunistic behaviour by the firms, the common denominator of equivalency that can express all round well-being through any rational means other than money, is awaiting to form an expression. Another way to look at the same problem is to relate organizational inertia to resist change, holding environmental and social commitments on the same pedestal as economic prosperity, though it also depends on infrastructure such as policies and legal and market arrangements that would reward firms with high environmental or social performance, even if at a cost to economic performance. This would need a rearrangement of ethical values and societal willingness to go beyond short-sighted causal motives to improve collective cognition towards universal well-being causal.

So, while firms could think of moving away from isolated environmental care to business models that reflect improved care for the environment in some cases, in the absence of standardized constructs to capture and enact firms' environmental conduct, economic thinking would still be the prevalent view and financial performance to be the sole judge of its achievement, which postulates the inevitability of environmental sanctions being closer to reality.

Methodological Developments in Environmental Management Accounting

INTRODUCTION

Environmental management accounting (EMA) is collective insights of scholars and practitioners, contributed to evolve management accounting to embrace environmental thinking. In this process, tools and techniques have been developed and improvised in theory and experimented through real-life case studies that were designed to gather information on materials and resources waste and diverted costs. This chapter covers how research and scholarship have been instrumental in developing new methodologies for capturing and developing information on environmental aspects of business activities. Additionally, in this chapter, we review how the cost of various aspects can be derived and be made a part of the information system. In this process, the divergence in valuation methods and the risks associated with assigning numerical values as 'costs' to the environmental aspects are explored. Some of these methods have been tested through case studies, while others are more at a conceptual level. In terms of EMA techniques, waste accounting and reporting develop information on waste by targeting expenditures incurred in generating and/or avoiding waste, while MFCA can compute the cost of waste by analysing the cost of ingredients and the resources used for their conversion. LCC builds the cost of environmental considerations into the asset acquisition or development process, while FCA covers the development of a costing methodology to include externalities along the chain of activities by breaching organizational boundaries. These costing methods and their comparative positioning vis-à-vis traditional ones are key to bringing environmental considerations into new methods and techniques.

EMA: A NEW ENVIRONMENTAL ACCOUNTING METHODOLOGY

EMA did not grow directly out of the subject of management accounting. It passed through the filters of conscious rationality and a series of questions and evolved through progressive elaboration of human understanding in this direction. Once the reality of the unsustainable practices of the industrial world was established, researchers questioned the economic viewpoint of accounting and its role in solving sustainability concerns. Though accounting claimed to have a sufficient armoury already to handle the treatment of environmentally sensitive costs, the generally accepted accounting policies did not step beyond the organizational boundary to address the bigger concern of environmental impacts. We saw two distinct trends contributing to advanced EMA. They were:

- Work around material and energy flows of organizations and look through the waste streams and save on costs by improving materials and resource utilization. These methods have gone down well with the controller groups in organizations due to the ease of using the accounting viewpoint (Kokubu and Kitada 2010), although needed an organization to establish a parallel information system and capture relevant data due to a distinct way of treating the material and energy flows, as we shall see later.
- On the other side, Life-cycle based methods offer a different set of tools for analysing data and reflect on environmental implications, it would require large data sets to perform the analyses, but which can be addressed through either subscription to a database or development of a custom-built in-house information base. Most of these decision-making tools attempted to incorporate costs (from within and outside the organizational boundary) to the extent feasible within the overall framework of these methodologies.

Some of these methods and techniques are being discussed in this chapter.

UNDSD WASTE ACCOUNTING METHODOLOGY

USEPA (1995) was the first agency to promulgate ECA and generate environmentally sensitive cost information. This was followed by the waste accounting methodology developed under the aegis of the UNDSD (2001). Waste accounting methodology is the first accounting-inspired methodology to use environmental cost drivers for identifying and classifying environmentally sensitive expenditures from the accounting ledgers. The method is simple to implement and has been piloted through several case studies. This waste accounting method generates post-operative statements reflecting the environmentally sensitive incomes and expenditures of a firm (Jasch 2003, 2006). However, such a restatement remains within the traditional boundary of financial accounting and does not incorporate environmental contingencies. Based on the expenditures incurred by organizations in creating and avoiding wastes that remain buried within the administrative overheads in financial books, suitable environmental drivers can be explored to identify and set up a causal link to these expenses. With the help of environmental cost drivers, expenses can be controlled and grouped under the cost categories of (a) material costs of waste, (b) non-product outputs (NPOs), (c) waste and emission costs, (d) prevention and other environmental costs, and (e) research and development costs (Jasch 2006). These cost categories represent expenditures being diverted from the organization's value chain. The basic formula to derive environmental costs is:

Total corporate environmental costs

= Costs of wasted materials + Costs of wasted capital
and labour (NPOs) + Cost of waste and emissions
+ Environmental protection costs (emission treatment and
pollution prevention) + Research and development costs   (8.1)

Several case studies across organizations have explored environmental accounting to elucidate the feasibility of defining a continuous reduction in environmental costs.[1] Besides the NPOs, the costs developed in this method would derive information from different accounting heads, whereas NPO needs information from cost accounting data to compute costs and allocate them to waste (Gale 2006). MFCA was one of the oft-used methods to generate the cost of NPOs from the production cycles. This is not to argue that the identification, separation and classification of environmentally sensitive costs and revenues from business transactions would reclassify recorded transactions into environmental sensitive expenditures and incomes and generate a post-hoc statement. Information generated through the analysis of accounted transactions would empower management with an alternative arrangement of costs. However, these methods are not designed to include costs that would be incurred by society—at present or in the future—to handle the negative aspects of waste and by-products (Jasch 2006; UNDSD, 2001). Subsequently, the International Federation of Accountants (IFAC) included externalities within the EMA framework by introducing the additional cost category of 'Less Tangible Costs' to cover environment-related externalities not available within corporate accounting records (IFAC 2005). This changed the definition of environment costs to:

Total corporate environmental costs

$$= \text{Costs of wasted materials} + \text{Costs of wasted capital and labour (NPOs)} + \text{Cost of waste and emissions} \\ + \text{Environmental protection costs (emission treatment and pollution prevention)} + \text{Research and development costs} \\ + \text{Less tangible costs} \qquad (8.2)$$

This development was an improvement compared to the earlier definition of EMA (UNDSD 2001) and broadened the definition of environmental costs by incorporating certain internal and external costs that were difficult to quantify, including externalities, and shifted the focus of EMA away from an eco-efficiency based outlook, at least in theory. However, a review of the literature from the later period revealed that

---

[1] For example, refer to case studies by Gale (2006), Hyrslová and Hájek (2005), Jasch (2003), Jasch (2006), Jasch and Danse (2005) and Staniskis and Stasiskiene (2006).

the externalities and less tangible costs did not always form a part of EMA scholarship (Jasch, Ayres and Bernaudat 2010; Laurinkevičiūtė, Kinderytė and Stasiškienė 2008; Lee 2011; Stasiškienė and Juškaitė 2007; and Viere, Schaltegger and von Enden 2007). In other cases, where its inclusion is witnessed, diverse perspectives are offered in the literature. Papaspyropoulos et al. (2012) have opined that the economic valuation of environmental aspects can be included as a part of less tangible costs and have used the indicators to map physical accounting of these aspects. Burritt, Herzig and Tadeo (2009) have factored in the potential revenue generation from certified emission reduction units (CERs) to evaluate investment in cleaner production, whereas Schaltegger, Viere and Zvezdov (2012) have avoided explicit monetization of externalities, but considered emissions in physical quantities while illustrating the EMA approach to cleaner production in beer manufacturing. Clearly, a lack of uniformity in integrating externalities and less tangible costs are visible within EMA scholarship.

MATERIAL FLOW COST ANALYSIS (MFCA)

MFCA was developed by the Institut für Management und Umwelt (IMU), Germany, as a technique to allocate costs to waste materials produced during manufacturing and processing activities, following the principle of process costing. In process costing, the cost of all the ingredients are allocated to the finished product, following the economic principle to transfer costs between by-products, joint products or co-products. MFCA differs in the apportionment of costs based on output quantities (including wastes), thereby treating waste as a joint product. As a result, using this technique, the cost of waste is assigned based on the cost of materials and resources that are being turned away from the value chain. Accordingly:

Material input + Starting inventory

= Finished goods + Ending inventory + Material loss   (8.3)

or, (Material input + Starting inventory) − (Finished goods + Ending inventory)
= Difference (material loss) or wastes   (8.4)

Following the principle of mass balance, MFCA could develop an alternative interpretation of waste by assigning economic importance to it (Burritt, Herzig and Tadeo 2009; Onishi, Kokubu and Nakajima 2009). Experiments conducted within Japanese industries using MFCA reported improved yields and process efficiencies (Nakajima 2011).

Cost of waste = Cost of finished goods (FG) inventory
  − Cost of inputs

$$= \text{FG quantity} \times [\text{Total input costs}/(\text{FG inventory} + \text{Waste quantity})] - [\Sigma(\text{raw materials (RM) quantity} \times \text{input costs}) + \text{Cost of resources}] \qquad (8.5)$$

However, MFCA did not alter the cost structure or incorporated costs that are contingent on and/or outside the boundaries of the firm. In that sense, the transactional boundary of MFCA remained within the economic realm of the business. Within EMA, MFCA ranks closest to the existing set of costing tools. This method was popularized in Japan by the combined efforts of the Ministry of Economy, Trade and Industry (METI) and the Ministry of Environment (MOE) (Kitada, Okaba and Kokubu 2009). Kokubu and Nakajima (2004) experimented with MFCA implementation repeatedly and provided an in-depth analysis of using EMA. The studies traced material flow quantities at the operational levels and recorded the waste as negative products. In places, MFCA was used as an extension of the existing ERP system, where cost data were simulated to generate a quantified waste value from different processes. Nakajima (2009) revisited the case studies on MFCA experiments in four of the major Japanese organizations as pilot studies a decade later, to offer insights on using the cost accounting tool that could help the management with the environmental cause by offering visibility of losses within the material chain, which is not available in the traditional costing methods.

Kitada, Okada and Kokubu (2009) compared the case findings from the earlier implementation of MFCA to the small and medium enterprises (SMEs) and came to the conclusion that this method helped in uncovering aspects of wastes that had been ignored earlier. In the case study on the Nihon Denkei Kagaku Co., Ltd., the company could identify the process deficiencies and improvise the processes to generate lower waste and improved quality levels. The study verified that SMEs are characterized by relatively weak negotiating position and fewer management resources, which could become constraints in successful implementation of MFCA. Still, it could help SMEs to improve resource productivities in a shorter time frame. Kokubu and Kitada (2010) have reasoned that MFCA helps management to look at operations in a way that is different from the traditional costing approaches, so it would be beneficial to adopt non-traditional thinking and take full advantage of it. Based on the adoption of MFCA in three organizations and the concepts from responsibility accounting, the authors illustrated that a different approach within MFCA might present conflicting situations for the decision makers if environmental decisions are weighed against economic benefits. These case studies demonstrated the ability of MFCA to improve EMA and the role of environmental accounting in decision-making. These studies have also examined the practical utility and operational aspects of MFCA implementation (mostly) within the manufacturing industry. The International Organization for Standardization (ISO 2010) has inducted MFCA as the ISO/DIS 14051 standard and introduced quality center (QC) as a unit of production, service or warehousing, within which material flows can be studied. The new ISO standard could be considered as being a step closer to connecting material flows with quality and manufacturing functions. However, MFCA is not designed to include waste if produced outside the mass balance, as in the case of emissions that do not form a part of the framework, not to mention the inapplicability of the framework to the services and projects industries (for example, tourism, airlines and EPC industries).

ENVIRONMENTALLY ENHANCED LIFE-CYCLE COSTING (E-LCC)

For environmentally sensitive LCC or E-LCC, the definition of cost is improvised to add the ones that would neutralize environmental impacts during manufacture, usage and disposal of products. The imputed costs are calculated on a 'voluntary basis' or on the basis of the 'polluter pays' principle (Gluch and Baumann 2004). The 'polluter pays' principle offers the commercial equivalent of a reconditioning for environmental damages where the organization only pays a token amount on a voluntary basis to legitimize its activities. Based on the success criteria in these projects, organizations can choose to act accordingly. Due to the need for exploded information, these methods might prove to be data and time sensitive. However, in case of unavailability of suitable data, proposals may still be worked out using suitable proxies. The impacts of proxies need the firms to weigh the size of approximations these might introduce. In addition, organizations can institute life-cycle tools by implementing business-relevant ISO standard(s) (ISO 14040-9).

$$LCC = \Sigma\ P_{n,r}(1+r_i)^{-t} \text{ (using hurdle rate principle)} \qquad (8.6)$$

where   P = payment,
        r_i = interest rate, depending on current environmental impact (low to high),
and,   t = time

or, $$LCC = \Sigma\ P_n(1-e)^t(1+i)^{-t} \text{ (using price rate principle)} \qquad (8.7)$$

where   P = payment,
        i = interest rate,
        e = escalation rate, and
        t = time.

Geissdoerfer et al. (2009) have debated whether LCC can be a standardized model for cost management. Viewing LCC as a non-standardized model, the authors postulated that it be used along with the TCO—a common model within specific types of decision making, for example, capital investments. However, the use of these models across industries is a matter of debate. Rebitzer et al. (2004) proposed to extend the usage of LCA to measure the social and environmental performance of products and services. The tool could calculate the impacts as part of the responsible product or service strategy and link CSR with corporate environmental management. For the purpose of establishing a methodological framework, a step-wise approach can be used to derive the impacts of individual CSR elements. Lindholm and Suomala (2007) have provided insights on developing the cost of products through the life stages by using life-cycle costing (LCC). In a case study of producing field guns from the Finnish Defence Forces (FDF), the product life-cycle costing exercise developed the cost profile of different life stages. It was proposed that deeper insights from the field could be used as feedback in the design and manufacturing of the next prototypes and could include the views of the customers as well. Although the exercise did not invoke externalities, it created sufficient room for them to be considered. From the review, it can be inferred that LCC can embed environmental implications as a part of its methodology. Environmentally sensitive LCC would depend on a considerable amount of data that might not be available within the organizational records. It might need a separate database to hold the information needed for it to use and derive decision options. Its sensitivity to discounting factors to derive the present cost of future environmental obligations is a matter yet to be settled. Still, the technique can sensitize the decision makers to environmental considerations and uncertainties that are a function of time.

FULL-COST ACCOUNTING (FCA)

The prevailing accounting framework limits the accounting process within the economic boundary of operations and covers operational costs of transactions by following the GAAP. This leaves difficult-to-measure costs, including social and environmental liabilities of economic activities, outside the accounting framework. While monetization of external costs is one end of the problem, building an objective interpretation for decision-making is another. FCA is an accounting technique that has been proposed in the literature to incorporate a *complete* range of costs, beyond what may be recognized in the books while following GAAP (IFAC 2005). Even though FCA can be considered as an ideological shift towards an 'inclusive costing' technique, it is yet to develop into a practical accounting tool. FCA is capable of including not only financial costs, but also hidden costs, contingent liabilities, intangibles and environmental costs (Bebbington et al. 2001). However, these costs are not always easily ascertainable, may vary with the scope of investigation and could reflect the interpretive bias of the economic agents (Herbohn, 2005). Despite some successful attempts by companies like Ontario Power Generation, BSO/Origin, PowerGen and others, setting out frameworks to capture external costs has proved to be time-consuming, tedious, inconsistent and fraught with methodological concerns (Antheaume, 2004).

Full cost of a cost object = Nominal costs
+ Contingencies + Intangible costs
+ Environmental costs
+ Hidden costs (if any)   (8.8)

Nominal costs include direct and indirect costs that the formal allocation process attributes to cost objects, whereas contingent costs are costs that depend on organizational exposure to certain risks, including penalties for non-conformance or non-compliance. Intangible and environmental costs form a part of the externalized liabilities that accrue, depending on how input materials have been procured and processed versus how waste and outputs are distributed in the market. Hidden costs can be a bucket or a placeholder to calculate costs that remain invisible due to different reasons, such as contractual misjudgements, misallocation of ownership to fix a burden, or costs indeterminable due to methodological failures. The issue with FCA is not only that the methodology has to calculate costs beyond the traditional bounds of arithmetical procedures, but the need to ascertain whether it is reflecting reality reliably and accurately for the decision-makers to rely on. Here, we need to include the boundary of considerations. The earlier examples of FCA were ideal but impractical ways to determine what could be the 'full' cost of an object and how much further the investigation needs to be carried out to pull out all the relevant costs. This raises the need for ascertaining the boundary to determine full costs instead of pushing for arithmetical accuracy.

In another advance, Bebbington, Brown and Frame (2007) have proposed the use of sustainability assessment models (SAMs) by combining FCA with the multiple-criteria decision analysis (MCDA) method for effective decision-making. With respect to the decision-making, MCDA acknowledges a plurality of values that leads the growing complexity of decisions within socio-political contexts. Instead of using CBA, which shows an over-reliance on monetization, the subjectivity of calculations, political influence (as a positivist method), distributional issues (with a focus on monetary goals, without worrying about how costs and benefits are distributed) and an over-reliance on experts, MCDA supports the dialogic approach to decision-making. Contrary to CBA, subjectivity is a recognized area in SAM and allows discussions amongst participants to increase objectivity in socially relevant concerns. Similarly, it covers distributional issues by including stakeholders without needing them to be experts. To counter unsustainable development, FCA could be one of the techniques to integrate environmental considerations with business decisions, where monetization can consider other economic methods such as the damage costs approach, shadow pricing and eco-assessment methods to monetize environmental losses by developing new techniques of cost estimation (Dascalu et al. 2008) when it comes to any types of costs other than what is already available through deterministic models. Another way to look at it is to capture the cost of activities that are a part of the same cost objective, even though the activities could be taking place within different zones of controllability, for example, the social costs of waste management that can be used as a proxy to estimate the amount of externalized liabilities.

ENVIRONMENTAL PERFORMANCE INDICATORS (EPIS)

The literature evidences that to measure the environmental performance of organizations, different parameters are to be set up as goals and their values should be tracked over time. In some research models, different environmental performance indicators (EPIs) have been formulated and studied as a part of 'eco-efficiency' (Ehrenfeld 2005), whereas in other cases, authors have proposed using a balance scorecard and improvised upon it with sustainability indicators (Laurinkevičiūtė, Kinderytė and Stasiškienė 2008). Indicators are defined in general as a means or mechanism to gather information about a phenomenon, where the significance of an indicator extends beyond the properties directly associated with the parameter values. For example, regional sustainability can be exemplified through the development of sustainable cities and a city sustainability index (CSI), which can be used to assess the cities' sustainability performance in relation to the impacts they has on human life and the environment, as opposed to their economic contributions (Mori and Christodoulou 2012).

Even though indicators have sometimes been vague in their import, they can still be indicative of the general direction of the underlying phenomenon. Accordingly, there have been several sustainability indices developed by different institutional agencies to cover sustainability at the regional levels and which can be applied to and compared across regions and countries in assessing the state of affairs. These indicators could be the indexed ones—for example, the living plant index (LPI), city development index (CDI), human development index (HDI), environmental sensitivity index (ESI), well-being assessment or well-being index (WI); or they might remain computational in nature—for example, an ecological footprint (EF), green national product, and others. These measures assess the social and environmental implications of economic developments at the regional levels and have been examined in detail in several research articles, including how these are derived and characterized as a part of policy choices. The studies have also been reflective of the lack of entirety that needs to be covered for them to signal sustainability or the lack thereof. Moreover, the measures have not always been based on a scientific approach of normalizing and weighting the individual or constituent index, including a scale that varies between intervals and ratio, and measured in monetized or physical units (Böhringer and Jochem 2007). Mayer (2008) provides a comparative view of the characteristics of some of these indices (Table 8.1).

In comparison to the organizational or product-based sustainability indicators (discussed later), the indicators in Table 8.1 are conceptualized to assess the progress of a region and, especially, to get a buy-in for policy choices. Elected leaders and decision makers need information to work through the political and economic choices and to pursue the balanced growth that the established frameworks should be ensuring. With the increase in awareness about sustainability, public officials at least would not be looking to get caught on the wrong side of public opinion. The literature has also indicated the role of local municipalities and their political agenda in driving the

Table 8.1 Comparative characterization of different regional indices

Index name	Aggregation method	No. of indicators	Distance to policy target?	Leakage detected?
Well-being	Average	36	Yes	No
Environmental sustainability	Average	76	No	Yes
Natural capital	Sum	7	Yes	No
Satellite-based sustainability	Ratio	2	No	Yes
Sparse Principal component analysis (PCA)	PCA	X[a]	No	No
Stochastic Impacts by Regression on Population, Affluence and Technology (STIRPAT)	Regression	11	No	No
Fisher information	System	X[a]	Possible	No
Sustainable economics, welfare, genuine progress	Accounting	X[a]	No	No
Genuine savings	Accounting	X[a]	Yes	No
Sustainable national income	Accounting	X[a]	Yes	Yes
Emergy analysis	Accounting	X[a]	Yes	Yes
Ecological footprint	Accounting	X[a]	Yes	Yes

Source: Mayer (2008).
[a] X—Depends on available data.

implementations of these programmes. The implementation of a local sustainability development index (SDI) covering 16 municipalities of the Local Agenda 21 (LA21) project in the Algarve region of Portugal has been discussed to show what needs to be a part of the local agenda and the implementation of sustainability index in measuring it (Ramos 2009). Even though scientific projects can develop different sets of indicators in a bid to capture the sustainability interests of a region, their adoption and further refinement as part of the accounting construct is a social process, which requires guiding and influencing from political and policy-related choices (Rametsteiner et al. 2011). As such, SDI creation at a regional level needs to be a 'norm creation' exercise that needs to be politically acceptable for the region to adhere to and pursue. At a minimum, this should introduce discussions in setting the political agenda and involve multiple stakeholders to form a democratic process in adapting to and agreeing on the indicators as a measure of the sustainability goals of a region and ensuing communication process among stakeholders (Shen et al. 2011).

While there are citizen-led (consensus-based) and expert-led (scientific) models to develop these indicators, there has been a lack of well-defined guidelines in terms of defining the sustainability of a given city or town. Turcu (2013) has used economic, social, environmental and governance themes to develop an integrative set of indicators by following a mixed-method approach (bringing together experts as well as citizen groups to improve the sustainability indicators) and used 170 indicators from 30 themes from five different published lists of indicators that are filtered and realigned using the inputs from experts and citizens. Shen et al. (2011), however, studied development practices through the development models of nine cities from around the world and compared these, as part of the international urban sustainability indicators list (IUSIL) that was developed by using a composite set of indicators from different models such as UN Habitat and others, and compared these to provide insights on the current practices and sustainability performance of these cities. The application side of the urban sustainability index reveals differences in practice (ground realities) and the resulting variations in terms of policies and preferences. The possible reasons for these variations include the intent of capturing different local interests within a standardized set of indicators, and the lack of reference that the standard indicators do not carry to correspond to the local needs. Another reason could be the difficulty in reflecting the local diversity through ratios and numbers that do not carry signals or meanings to reflect the uniqueness of the concerned region. In addition, some of the decisions or policy choices could have long-term impacts on the growth and resource map of the regions and, accordingly, in the absence of much-needed causality information, would lead to difficulties in calibrating the impacts on a long-term basis. Another possibility would be to monitor the indicators on regular basis and develop new models that would improve 'sustainability', which is hardly defined as the end objective.

At the level of the industrial sector, the growth in literature supports development of indicators that are attuned to the activities of the sector—the building construction sector is a good example to reflect this trend. The building construction sector has developed standardized assessment frameworks such as Leadership in Energy and Environmental Design (LEED) in the US, which is gaining popularity in terms of its applicability and considered as the gold standard in this sector. However, there are other assessment frameworks as well, such as the Building Research Establishment Environment Assessment Method (BREEAM) in the United Kingdom (UK), the Green Building Tool (GBTool) in Canada and the Comprehensive Assessment System for Built Environment Efficiency (CASBEE) in Japan. The advantage of using one framework versus another could be explored by developing metrics for a side-by-side comparison of their comprehensiveness, design guidelines, signalling and communication method. Comprehensiveness in this case might cover the type of building, materials, locations and other diversities of construction, whereas design guidelines offer a way to adapt to the available conditions; signalling and communication methods pertain to the target and its achievement. Having said that, I do not mean that the frameworks are free from theoretical and practical shortcomings—for example, ambiguity in weighting factors, lack of utility to the owners, or failure to reflect advancements in knowledge and learning from their adaptation across regions—which need to be rectified and improved upon (Kajikawa, Inoue and Goh 2011). The point is the incompleteness of the different frameworks is progressive elaboration of our contemporary understanding of the underlying complexities. So any expectation of an all-encompassing solution would only be a utopian hope.

Environmental performance indicators at the firm level

In comparison, environmental indicators at the corporate level are a way to reflect on how organizations are working towards improving their efficiency with respect to materials, energy and resources used in the products, processes and services that the firms offer. This could include instituting different types of eco-labelling programmes to validate environmental sustainability of some sort, in order to appeal to as well as inform customers of firms' environmental and social proactiveness—for example, the various Energy Star ratings to reflect energy consumption, the Green Star measuring sustainability of buildings in Australia, the Home Energy Rating System (HERS) in the US, the Blue Angel in Germany or the Nordic Swan of Sweden, or others being promoted by different institutional and regional agencies to encourage sustainable behaviour and consumerism (Boström and Klintman 2008). However, the promulgation and use of different indicators reflect a lack of standardization at the international level and the commercialization of a consumption pattern that is based on the current trend of appealing to the altruistic needs of the consumers, satisfying some sort of self-gratifying need to be eco-sensitive and contributing to sustainability, which is becoming a common marketing practice across industries. Moreover, the traditional forms of the driving values as a part of human life are increasingly being replaced by material well-being, which the marketing channels use to promote one or the other form of consumerism even while articulating an appeal for sustainability (Dahl 2012). Stegall (2006) explains that sustainable design of products and services should not be limited to a specific segment of its life cycle (say, manufacturing or maintenance), and instead needs to develop life-cycle thinking, including post-disposal measures, which has hardly been fundamental to eco-labelling programmes.

In this context, EPIs at the firm level can only facilitate a comparison of certain parameters or ratios at a given point of time, which can over time help organizations to devise strategies to spot 'worsening' ones and initiate discussion on reversing or neutralizing these trends. Although scholarly opinion is divided regarding the benefits of using performance per unit of environmental impact (eco-efficiency: Braungart, McDonough and Bollinger 2007) versus environmental impacts per unit of organizational activity (eco-intensity), the idea behind the deployment of ratios is to improve processes, products and methodologies in generating outputs that could ultimately improve the environmental *bads* (Ehrenfeld 2005). EPIs can be adopted or

Table 8.2 Environmental performance indicators (Examples)

Resources/ Output	Eco-efficiency indicators	Eco-intensity indicators
Energy	1. Total mass of finished products (kg)/Energy consumed (kWh) 2. Total mass of waste (kg)/Energy consumed (kWh)	1. Energy consumed (kWh)/Total mass of finished products (kg) 2. Energy consumed (kWh)/Total waste of mass (kg)
Materials	1. Total mass of FG (kg)/Total mass of waste (kg) 2. Total FG value (currency)/Total mass of waste (kg)	1. Total mass of waste (kg)/Total mass of FG (kg) 2. Total mass of waste (kg)/Total FG value (currency)
Emissions	1. Total mass of finished products (kg)/Total GHG emission (tCO$_2$e) 2. Total FG value (currency)/Emission from energy consumption (tCO$_2$e)	1. Total GHG emission (tCO$_2$e)/Total mass of finished products (kg) 2. Emission from energy consumption (tCO$_2$e)/Total FG value (currency)
Waste	1. Total sales (currency)/Total mass of waste (kg) 2. Total production (t)/Total mass of waste (kg)	1. Total mass of waste (kg)/Total sales (currency) 2. Total mass of waste (kg)/Total production (t)

Source: Author.

Notes:
FG—Finished goods.
GHG—Greenhouse gases.
tCO2e—Tons of CO$_2$ equivalent.
MT or t—Metric tons.

designed as shown in Table 8.2. These and other such EPIs can be institutionalized and benchmarked for industries, processes and/or products, and could be implemented as part of EMS for control and reporting purposes.

$$\text{Eco-efficiency} = \frac{\text{Financial or physical performance}}{\text{Environmental impacts}} \quad (8.9)$$

$$\text{Eco-intensity} = \frac{\text{Environmental impacts}}{\text{Financial or physical performance}} \quad (8.10)$$

It would be pertinent to point out here that EPIs could be developed using internal (financial and environmental performance measures) as well as external data sources (KLD ratings, S&P green index, sustainability metrics indicator, ISO 14031, and others). However, it is up to the management to ask for measures that would help improve the performance of firms. The World Business Council for Sustainable Development (WBCSD) defined eco-efficiency as the performance level to deliver competitively priced goods and services that would satisfy human needs and bring quality of life, while progressively reducing ecological impacts and resource intensity in line with the earth's estimated carrying capacity (WBCSD 2000a). Organizational performance as well as its environmental impacts can be measured in diverse terms and accordingly multiple sets of indicators can be derived (Seiler-Hausmann, Liedtke and Weizsacker 2004). The trends of these indicators could be used to track organizational or industrial performance over time.

Scholars have not always been optimistic about eco-efficiency as a possible contributor towards sustainability. The lack of standardization could turn a comparison of the same ratio between competing firms somewhat asymmetrical. The EPIs, being measured as ratios of a mathematical nature, completely hide certain impact areas such as technology, quality and overall consideration towards a cradle-to-grave life cycle. Also, the ratio might show indifference towards relative consumption levels of resources or degradation due to the proportionate changes in the measures that could offset a change in the performance levels. This would result in nullifying directional changes in the measures, neutralizing the effects of performance on the environment (Gray and Bebbington 2001). Despite these shortcomings, eco-efficiency has been found to be relatively easy to compute and interpret. Also, industries can design their own measures that can be standardized as part of industry practices (IFAC 2005; WBCSD 2000b).

ARE WE THERE YET?

This chapter has reviewed methodological developments within EMA and explored different techniques such as MFCA, LCC and FCA.

However, each of these methods has its own take in terms of deriving what could be the cost of environmental aspects (*not* impacts):

- Waste accounting and reporting identifies the cost of environmental aspects as being equivalent to the expenditure incurred on waste, its prevention, treatment or disposal. In this way, it promotes the notion that firms can control waste by controlling these expenditures. In other words, the economic incentives of reduced costs and an improved bottom line could become driving factors in reducing waste, but are no better than aiming for low-hanging fruits.
- MFCA, in contrast, defines the cost of waste through its recipe or the ingredients that form a part of the bill of materials and resources. This pushes for reducing the waste as it accumulates costs due to materials and resources, followed by costs of operations or conversions, and ultimately the costs incurred in the disposal. Scientifically accurate and closer to reality in terms of how waste is formed, this would still be depending on the formula of mass balance formula to define what waste is. Moreover, the idea is to control costs associated with waste and reduce the levels of waste generation.

Accordingly, environmental costs in both these cases are controlling mechanisms to improve the usage of materials and resources and reduce waste. Also, both the methods are restricted to material waste.

- In comparison, FCA allows development of a cost function based on the chain of events that might stretch beyond organizational boundaries, capturing costs at every level to offer insights into the environmental impacts; it is easier said than done. Exploring the lifecycle and impacts beyond organizational boundaries would depend on time, cost, and efforts, an organization is ready to invest, including the willingness to pursue it for the sake of improving scientific accuracy of data and findings.
- Meanwhile, LCC demonstrates a capability to consider environmental impacts during and after the life of products and/or services and develops a cost function to derive the present value of investments.

So FCA and LCC can explore costs beyond times and boundaries and expand cost formulations. These developments are beneficial in building the knowledge base and for environmentally conscious decision-making.

- On the contrary, eco-efficiency (using EPI) represents a business-oriented approach to evaluate organizational performance by comparing the physical performance with the environmental aspects generated, or vice versa, a simpler yardstick for organizations to follow in order to improve operations in a way that would reduce environmental impacts per unit of output or per dollar value.

Even though the literature has advocated standardizing these measures within and across industries, so that firm-level performances can become easily comparable, the industry is far from embracing them in practice (other than MFCA, which is instituted as ISO 14051). This establishes the view that the ongoing discussions on the subject are far from over, including the one exploring the difficulties of firms in adopting, measuring and reporting their environmental performance. It is important here to acknowledge that the environmental performance of an organization is not an isolated incident in its history. Within the systemic perspective, it is the overall ability of people, processes, the operating environment, systemic environmental and competitive pressures, and improved accountability that the firms have to deal with to address these concerns.

Advances in Other Environmental Frameworks

INTRODUCTION

Contemporary developments within the domains of cost and management accounting have supported corporate environmental considerations in one way or another. A review of the literature also indicates the development of and experimentation with other pro-environmental frameworks to offer niche solutions, covering specific environmental considerations in businesses. This chapter is a review of the advancements found in the literature on carbon accounting, GHG accounting, the Carbon Disclosure Project (CDP) and sustainability reporting. These diverse frameworks are being instituted by different international bodies. While carbon accounting has relevance to the Annex I countries due to the legitimization of the Kyoto Protocol, GHG accounting addresses the inventorization of GHGs or Kyoto gases that a firm produces and should be responsible for. In comparison, sustainability reporting involves the development of a voluntary reporting process for the economic, environmental and social performances of the organization in line with a triple bottom line (TBL) perspective that firms might adopt; although it is still far from developing a common language of expression, these developments, however, relate to the emerging trends so as to develop a better visualization of environmental considerations and to align these within the overall scope of corporate green accounting.

GREENHOUSE GAS (GHG) ACCOUNTING

GHG accounting is an independent process of inventorizing the greenhouse gases associated with the organizational existence. GHG accounting has been standardized to cover the GHG-related risks of industries through multiple reporting standards issued by the World Resources Institute (WRI) and the WBCSD (WBCSD and WRI 2004). Although industrial activities might generate different types of gases, the GHG accounting standards apply to six gases—carbon dioxide (CO_2), methane (CH_4), nitrous oxide (N_2O), hydrofluorocarbons (HFCs), perfluorocarbons (PFCs) and sulphur hexafluoride (SF_6)—that are covered by the Kyoto Protocol, measured in the common unit of equivalent tonnage of CO_2 (tCO_2e). However, it is up to the firms to leverage the same mechanism to account for other (non-Kyoto) gases such as chloroflurocarbons (CFCs), the oxides of nitrogen and sulphur (NO_x and SO_x, as applicable) and various other fluorinated gases (F-gases) within the applicable scope and build or report an extended emission inventory (ibid.). TBL reporting indicators—EN16 to EN18—recommend that reporting entities follow corporate GHG reporting standards to report on GHG emissions (GRI 2006).

GHG accounting involves the identification of emission scope followed by categorization and calculation of GHGs that are released to the environment by business activities. Scope 1 emission is associated with the generation, transmission and distribution of electricity and other energy sources owned and operated by firms within their ownership boundary. Scope 2 emission is attributable to electricity consumption, sourced from external producers and distributors. It is pertinent to mention here that Scope 2 GHGs can relate to the actual consumption of energy of a process only if measurement instruments (e.g., energy meters) are installed to capture actual consumption. Otherwise, GHG generation would have to depend on the total energy input and would need subsequent allocation to processes to generate emission profiles. Scope 3 emission is classified as emission associated with all activities that do not belong to the other two categories (e.g., purchase, business travel, transportation of goods and services, customer services, and so on). The GHG standard for corporate reporting serves as a basis for business organizations to develop an emission inventory (for Kyoto and non-Kyoto gases), generate information to monitor and improve performances, and participate in voluntary or mandatory compliance and reporting.

The discussion on GHGs as a part of green accounting is relevant to emphasize that emissions are a part of the waste chain and should be within the corporate environmental accounting framework. With increasingly stringent emission norms, organizations would have to manage their carbon-related business risks better (Schaltegger and Csutora 2012). Huang et al. (2009) have discussed the use of input-output analysis (IOA) for analysing GHG across supply chains and the boundary for cut-off or threshold for emissions could be used for marginally costly resources. IOA can also be used as a screening tool to capture emissions from the supply chain and to derive anticipated life-cycle emissions. However, decision makers would need to know the complete footprints to take business decisions and would need to consider sectoral averages and potential sales demands to aggregate them and develop a reduced carbon profile of products and services. It would be difficult for corporate organizations to build a response towards rising emission challenges due to incomplete emission profile of supply chains and at sectoral levels. Development of GHG accounting and related research is confined to developing emission profiles for products and processes at the gross levels by developing inventories of GHGs. However, its systemic integration within management accounting to help management take corrective steps is being researched (Schaltegger and Csutora 2012). Burritt, Schaltegger and Zvezdov (2011) have reflected on the carbon management practices of some of the leading German companies to define what could be considered as 'carbon management accounting' (CMA). Defining it as a component of the sustainability accounting framework, the authors have explained how it could help organizations approach carbon management with physical as well as financial data that would be needed to participate in emissions trading, saving energy and gaining a market advantage. However, such integrations are yet to mature.

The Carbon Disclosure Project (CDP)

Kolk, Levy and Pinkse (2008) have studied the organizational responses towards global climate change through the study of CDP reports which have been initiated under an autonomous body that invites businesses to join a voluntary reporting process, similar to that of the GRI by the Coalition for Environmentally Responsible Economies (CERES). The GRI seeks reporting from constituent members on their sustainability performance based on the TBL approach. The CDP invites organizations, including institutional investors, to join it and report their carbon management and disclosure following the CDP5 questionnaire. While there could be future implications for linking voluntary carbon disclosure with carbon trading, its reporting through the CDP is still a matter of political debate. The CDP can be considered a form of corporate governance in which civil-society actors employ the disclosure mechanism to exert pressure on businesses to establish and comply with environmental and social norms. Due to the broad nature of the questions and subjectivity of the answers, there is apparently only a low possibility of generating quantitative information from the submitted data. However, the institutionalization process should be able to rely on the 'commensuration' to establish a valid logic or causal chain of events to convert qualitative information into quantitative one.

EMISSIONS AND CARBON ACCOUNTING

The climatic changes due to the rising levels of carbon dioxide and other GHGs is believed to create future problems such as lack of access to clean water, supply chain concerns due to weather and infrastructural strains, market risks due to the rise in sea levels, changes in customer needs due to unpredictable demands, and country-specific risks due to political and security conditions (Schultz and Williamson 2005). Since the government and supranational bodies have responded to the threat of global climate change with responses such as developing awareness, energy-efficiency measures, discussion on carbon impacts on consumer choices, and economic responses to global climate change, the greenhouse effect can be considered as real and threatening (Bebbington and Larrinaga-González 2008). Historically, the atmospheric concentration of CO_2 has stayed between the levels of 180 and 300 parts per million (ppm) for over 600,000 years. However, the Industrial Revolution led to a sharp rise in carbon dioxide concentration by the end of the last century and a further rise in the concentration could push the surface temperature up by two degrees Celsius (Carter et al. 2007), which has been considered by scientists as the threshold to the altered behaviour of the environmental system, producing damaging effects including the runaway impacts of weather changes (Bebbington and Larrinaga-González 2008; Schultz and Williamson 2005).

Since the institution of the Kyoto Protocol, the EU has issued emission quotas under a cap-and-trade scheme to approximately 12,000 industrial facilities within the EU and imposed limits on the carbon dioxide to be released by these firms. The total quota issued is less than the expected release of carbon dioxide, and hoped the scheme to stimulate organizations to find cost-effective solutions for pollution abatement. At the same time, the carbon allowance is traded as per EU Emissions Trading Scheme (ETS), where the shortage of carbon permits is expected to push the level of prices of carbon permits further. Considering the impact of an emissions quota on businesses, opportunities have to be discovered and converted into successful business propositions, turning carbon credits into profit. This would also need the firms to assess carbon exposure, develop competitive initiatives and rethink options to mitigate carbon risks (Rosendahl and Strand 2011).

The Kyoto Protocol also introduced schemes such as clean development mechanism (CDM) and joint implementations (JI) to balance the demands of emissions in Annex-B countries and incentivize them to reduce their overall emission. One of the ways to achieve this is to shift operations to non-Annex B countries (mostly developing countries). Non-Annex B countries are expected to secure projects and programmes that would reduce their GHG emissions as well as contribute to the regional development agenda. However, CDM is an offset mechanism and its objective is not to reduce overall global GHG emissions. Also, during the transfer of projects from carbon-restricted countries to non-Annex B countries, the leakages occurring within the processes due to the differential prices of the fossil fuels would go unnoticed. With the increase in numbers of CDM projects in non-Annex B countries, without any cap on emissions, consumption of fossil fuel would change its demand-and-supply pattern and contribute to the leakage (Rosendahl and Strand 2011). The Paris climate talks or the Conference of the Parties (COP21) in December 2015 legitimized the agreement of the countries to bring their levels of GHGs below 2005 levels by 2050 so as to maintain the temperature rise at sea level within two degrees Celsius of the present temperature.

Tol (2005) has surveyed the relevant literature to form a probability density function from the 103 estimates of marginal cost damage from carbon dioxide published in 28 studies and asserted that the best guess is around USD 50 per tCO_2e, even though the estimates depend on the discounting rate and aggregation of monetized impacts across countries. In comparison, the going rate of carbon in the year 2012–2013 was hovering around EUR 4–5 per tCO_2e in the ETS market and failed to match the earlier estimates, highlighting the controversial and uncertain nature of a market-based solution. However, the situation is fast changing, and in 2018, the prices of carbon have been steadily increasing, hovering around €25+ per tCO_2e and anticipated to move to €30 per tCO_2e, but this is still not close to the earlier estimates (Bloomberg 2018). Still, the COP21 agreement provides a new opportunity to firms to develop mechanisms benefiting from policy intents. However, as a knowledge gap exists in analysing the impacts of climate change, differences due to policy goals, growth models and ground realities could result in unpredictable outcomes. Moreover, not all impacts are well-understood, neither the nature of vulnerabilities that various regions contribute to the global estimates.

Accounting implications of carbon trade

Ratnatunga and Balachandran (2009) have offered a pragmatic approach for the incorporation of carbon accounting within the business accounting framework. Considering international emission trading to be an accepted reality in the future, it would be essential to account for the transactions involving emissions (including futures, assigned accounting units, Kyoto units, certified emission rights, and others), but it would also need improved accounting structures to incorporate emission trading as a part of overall business. Detailing the provisions of the Kyoto Protocol, Bebbington and Larrinaga-González (2008) have referred to the mandatory cap-and-trade system for Annex B countries within the commitment periods, where companies might participate to purchase and sell CERs in the voluntary market, as this could be an economic stimulus for companies to move towards low-emission alternatives. From the accounting perspective, the organizations would need to match their actual emissions annually with carbon allowances (available through a national allowance) plus the CERs and surrender these to the national registry. Violating the cap would lead to financial implications in terms of fines as well as the purchase of balance CERs from market at prevailing rates. The EU has set the penalty rate at €100 per European Union Allowance (EUA) while the market traded at €30 per EUA (Bebbington and Larrinaga-González 2008). However, this poses difficulties in deciding how to handle the grandfathered emission allowances and offsetting them against physical emission reductions achieved through clean production mechanisms versus arriving at the rightful closing value of the CERs.

At the same time, the correct procedure to measure carbon credits is another concern facing academics and practitioners, and there is no consensus regarding the method of valuation and uniformity of procedures. Bebbington and Larrinaga-González (2008) suggest that the recognition of assets and liabilities and their reporting might follow a net approach, where the allowances are considered to be granted at nil value while the obligation is recognized at the carrying value. There have been accounting implications all right, but no amicable settlement is within sight. It is believed that the accounting of global climate change would need a change in the conventional accounting system to cover the associated risks and assist decision makers. Since global climate change implications for countries are varied and no long-term solution is available yet, it becomes meaningful for the companies to manage regulatory risks and devise policy instruments at the supranational levels (ibid). In comparison, Machado, Lima and Filho (2011) have focused on the use of the IAS, such as IAS 20 (accounting for government grants and disclosure of government assistance), IAS 37 (provisions, contingent liabilities and contingent assets) and IAS 38 (intangible assets) for carbon trading. For the valuation of carbon credits, IAS 38 can be used by considering emission rights as intangible assets, while keeping in mind the differences in timing and valuation of these rights. If the permissions are allocated by the government at a value lesser than the fair market value, the differences could be recognized as per IAS 20, whereas any liability associated with the CERs or penalty should be accounted as per IAS 37, the authors suggest.

Ratnatunga, Jones and Balachandran (2011) have used the asset capability framework to develop the valuation and report carbon emissions. Terming it as an environmental capability enhancing asset (ECEA), they have recognized two types of carbon credits: (a) the ones issued by the government and (b) another kind generated due to internal capability enhancement. However, accounting for these would mean: (a) valuation and reporting of carbon credits, (b) valuation and reporting of intangible assets and (c) deciding how the organization is going to meet its environmental and social responsibilities. The accounting issues can be further related to: (a) the valuation of permits, (b) the valuation of liability over time and (c) the recording of grants. As carbon credits are intangible, the accounting process should take care of carbon offsets as well as the associated financial value. Another way would be to value carbon credits based on its prevailing market rate and not follow the historical cost to represent fair value accounting. The variations in the transactional values can be recorded using 'capability' states rather than the historical costs. For internal control purposes, Ratnatunga and Balachandran (2009) have proposed the use of product miles as a proxy for units of carbon emission generated due to product transport, thus accounting for the impacts of product movement on the environment. The third option would be to improve strategic cost and management accounting to provide a suitable structure. Clearly, carbon accounting is struggling to handle the aspects, set-off, carbon credits and associated financials in a seamless manner, more so because the current capabilities of the accounting frameworks are yet to provide for a mechanism that can simultaneously evaluate and account for economic and environmental views of transactions—an area central to this discussion.

SUSTAINABILITY REPORTING AND THE GLOBAL REPORTING INITIATIVE (GRI)

The GRI originated in 1997 out of CERES and it was established as a permanent and independent international body in 2002, with a multi-stakeholder governance structure, supported by the United Nations Environment Programme (UNEP). The reporting standards of the GRI follow a standard disclosure profile and allows the organizations to report on their strategy and analysis, organizational profile, reporting parameters and governance aspects. In addition, the report has six categories and each category includes various relevant aspects. The reporting categories are economic performance, environmental performance, human rights, labour practices, society, and product responsibility. These categories cover reporting of different aspects and divided into core and optional indicators. GRI indicators are derived from practitioners' perspective and are categorised into three sections and each section is further collected into various aspects. The reporting is voluntary and has scope for external assurance. The reporting levels are divided into A, B and C, with a decreasing number of mandatory indicators to be reported corresponding to each level. More details about the GRI and its reporting indicators can be found on its website (www.globalreporting.org). The benefit of using the GRI reporting option is its standardized format and prescribed methodology to calculate individual indicators. Although the reporting format encourages qualitative and quantitative analysis based on the reporting area, the platform is yet to publish any consolidated findings based on the individual reports submitted so far.

From the academic angle, a few important aspects may be noted:

- This reporting framework has been instituted by accounting practitioners for the industries and firms to participate in. This is advancement over the usually qualitative nature of CSRs and brings in some element of objectivity.
- The theoretical bases to calculate different aspects and their reporting have not passed a test for academic rigour, although the GRI recommends the firms to follow certain standardized methodologies.

- It is purely voluntary for firms to follow GRI reporting standards and up to the participants to decide which level of reporting they would like to do, so it is mostly peer pressure that is assumed to drive the improvements.
- Improvements in the internal workings of the firms and correspondence to whatever is being reported is left to the firms to decide, although assurance guidelines are also available, should a firm wish to turn in assured reports.
- Whether this would improve the internal working of the firms or environmental proactiveness remains an open question.

SILVER LINING IN INCOHERENT NEW DEVELOPMENTS

This chapter has covered the development of pro-environmental frameworks that have evolved in parallel to EMA. For example, CMA is proposed as an umbrella framework to institute systematic investigation of GHG-related impacts on decision-making. However, its interactions with the EMA framework are yet to evolve. Similarly, GHG accounting developed as a standalone construct to inventorize Kyoto gases. While it could be argued that the effectiveness of the EMA framework would improve if it were based on the environmental performance of the organization, covering waste as well as emissions, in the absence of any interaction between EMA, CMA and GHG accounting, any form of emission accounting remains only a partial solution. In addition, the absence of accounting standards covering carbon trade and accounting remains a problem. Still, there are a few relevant takeaways in the proposed scheme of things. The advancement of GHG accounting developed the concept of environmental assets (to reflect waste and emissions) while sequestration reflected the liquidation of the assets.

Even though GHG accounting took care not to indict businesses with the environmental assets they produce, it connected firms with the emissions they release to the environment that roughly translates to the abandoned assets of the firms with firms retaining quasi-ownership rights over these. Second, it brought home the concept of accounting for environmental assets and their liquidation due to sequestration by using the stock-and-flow method that can be temporally traced, similar to inventory tracking. However, GHG and its accounting remained independent of the normative accounting framework due to the absence of a connecting argument that can establish a commonality between these two systems. Due to the quantitative nature of GHG, which can very much be derived using a computational solution, the need for a separate accounting was never brought up in scholarship. Other than GHG accounting, no other framework tested or developed constructs that considered temporality as a requirement. Third, most of these methods remained occupied with physical quantification and found monetizing of impacts challenging due to insufficient theoretical contribution. All these frameworks acknowledged the multitude of measurement units and undefined monetary equivalency. Also, limited efforts were seen to extend the methodological considerations beyond organizational boundaries and derive a scientific basis to calculate the cost of externalities. In that sense, the efforts concentrated on reducing the quantity of the environmental aspects and corresponding impacts.

For an organization to consider its environmental performance seriously, it would need a construct that offers a 360-degree view of the various aspects, including credits, offsets and the associated financials on waste. Such a construct would need uniformity in methodology to quantify and monetize aspects and remove interpretive bias. In addition, it would have to concern itself with a multitude of accounting principles (formal versus informal types of accounting principles), measurement differences (financial versus non-financial or mixed units) and content of information (information of aspects vs impacts). Transparency and authenticity of information are vital and need to be taken care of by the accounting information system (AIS). In the next section, the journey we undertake is to learn from this discussion and evolve theoretical foundations for a plausible breakthrough.

Environmental Accounting: A Dimensional View of Accounting

Environmental Accounting

Connecting Critical and Normative Theories of Accounting

INTRODUCTION

Business organizations form a major pillar of societal existence and they need an accounting solution to help them better manage, report and control externalities (proactive approach), instead of waiting for policy enforcements to act responsibly (reactive approach). From the previous section, we understood that SEA evolved as a philosophy to meet the need for firms to be accountable for environmental and social externalities, while EMA provided the methodological support to influence internal decision-making processes. However, these efforts did not push the accounting processes to change significantly. The prevailing methodologies are struggling to help organizations go beyond the obvious, failing to integrate the 'external' costs of waste that society would have to incur at present or in the future and failing to mitigate the negative effects. The inadequate coverage of poorly defined ownership of waste, lack of corporate accountability towards social and environmental concerns, and inheritance of economic theories in accounting have contributed to shape corporate accounting system that ignores externalities. Also, stakeholders' concerns about the environmental performance of organizations create a tension for market-oriented performance reporting, where the firms need the capabilities to handle environmental performance alongside financial ones. This mandates that the accounting solution (whatever it might be) must collaborate with environmental methodologies (emission accounting, life-cycle methodologies, flow cost accounting and GHG accounting) to offer transactional transparency for the environment–firm exchange, help firms take corrective steps (e.g., from a choice of cleaner technologies) and contribute towards environmental sustainability (e.g., by way of resource usage, a cradle-to-cradle cycle, or a no-waste policy) without losing the temporality that is central to any accounting system.

THE ENVIRONMENT AND ECONOMICS: TWO DISCONNECTED PARADIGMS

Economics is all about employing scarce resources to generate benefits for the human world, and the participating enterprises are rewarded with returns for its efforts. Here, the concept of 'value' represents the utility value of goods and services, but evaluated using the market worth of the outputs that the transformation process has generated. At the same time, the suitability of using one transformation process versus another—or for that matter, policy formulation to promote one path of transformation versus another—is not dependent on the comprehensive assessment of all the impacts that the chosen transformation process would produce. The impacts of the transformation processes resulting in a degraded environment is obvious, but not many scholarly contributions are available to generalize such impacts or offer insights about countering these. This could be due to (a) human knowledge of the complete environmental impact of an anthropocentric event being incomplete and (b) not having any language other than the monetary scale to translate the results of these impacts. The limitations result in a non-convergence of environmental questions and economic choices, which has to do with the absence of a measurement scheme, with the human perspective reducing everything to a single denomination and a myopic outlook that does not venture beyond economic interests. Accordingly, there is a need to define 'value for whom'?, which overlaps with anthropocentric views. It would be too soon to say that humans have converged these two different realms beyond an illusion. As the normative view of accounting is ingrained in the economic views of the firm, bringing environmental impacts within the accounting process would not change the outcomes unless we can somehow ensure the convergence of the two different paradigms at a common point and experiment with the outcomes that it leads to. Beyond this, the challenge is to highlight the importance of human activities and their impacts on the ecosphere, without which human endeavours might end up shortchanging the ecosphere's welfare for anthropocentric gains. Accordingly, my task is two-fold here:

- first, to connect these two paradigms and their epistemological boundaries (critical theory versus the normative view) and explore whether such a convergence can be sustained (through epistemological advancement), and
- second, to establish that it is possible to use the accounting construct to capture, translate and record the environmental impacts of business activities (through methodological improvements).

This, I hope, would translate into the long- and short-term implications of our decisions and explain how remaining blind to these implications is dangerous to our own interest (the ontological view).

WHAT ARE WE MISSING?

To recapitulate, accounting innovations have contributed to the legitimacy claims of firms and their efforts to lower waste, improve profitability and gain mileage in the short term. In contrast, SEA theories have legitimized the use of corporate sustainability reports and other dissemination channels to institutionalize firms' accountability to stakeholders, in the hope that this would be instrumental to changing the behaviour of the firms to be socially and environmentally responsible. However, the approach of using reporting as a solution could only be a blind guess because of the challenge to ensure neutrality of the reporting processes. While the shift in focus from 'accounting' (taking the normative view) to 'accountability' (with an ethical focus) is a good beginning, advances within the normative approach have to be backed up with accounting innovations and the enactment of new developments. For example, if enhanced environmental stewardship is perceived and accepted as a part of the firms' accountability, how would accounting functions validate the claims that the firms are working towards new goals? Interestingly, the aim to improve accounting functions to bear the responsibility towards society and nature has not challenged the core accounting constructs (normatively speaking) so far. This myopic approach could be due to multiple reasons—for example:

- Unbreakable tie with the economic paradigm that severely limits the capabilities of core accounting constructs to accommodate the new demands of changing business responsibilities, which cannot be aligned to economic aspects. For example, FCA was used by some of the companies to develop an all-inclusive framework to capture costs along the entire life cycle of their business activities and account for these costs, but it did not prove useful enough.
- Methodological complexities and the poorly defined causality of any such innovation, as in case of FCA, can make it an unviable choice. Thus, only a handful of companies could develop detailed accounts of extended costs (Bebbington et al. 2001; Herbohn 2005; and Antheaume 2004). In these experiments, limitations emerged in trying to define extended costs, not to mention the sensitivity of the costs that resulted in changing the values significantly, resulting in loss of decision makers' trust.
- Another reason could be the perceived equivalence of the accounting language with that of the accounting framework in use, as if there is only one way in which the accounting language can be expressed, with any other form of accounting or the possibility of its enactment can only be achieved by extending financial accounting, and that view remains a barrier to the scholarship. Although not entirely related, a similar predicament was expressed by Mattessich (1971) regarding reviewing the theoretical and foundational problems of the accounting sciences where solutions to accounting challenges are sought by concentrating on specific narrow areas (such as financial accounting in isolation), without searching for an overall theory that might comprehend all other areas of accounting and information system.
- A third reason could be the successful persistence of the basic *(financial)* accounting construct to evaluate businesses in market-based economies that carry a fear of disruptive changes and the non-reconciliatory ruptures in the overall construct that these

changes might lead to, which can result in destabilizing economic interests beyond repair. This might as well be a good reason to avoid looking for an overarching theory in accounting.

I earnestly believe that time is of the essence as we search for a fundamental design or a design change that will abstract the generic accounting process and its enactment—where a critical view can become a part of the normative accounting world. The next section explores this further.

HOW FIRMS MEASURE AND ACCOUNT FOR THEIR SUSTAINABILITY DRIVES

Due to the lack of convergence that plagues the micro-level application of sustainability within the extant literature, the first part of the problem is to develop a boundary of accountability for firms to operationalize. The basics of control theory suggest determining a sphere of influence for this purpose. For example, although firms can be held accountable for emissions, their sphere of control is limited to direct and indirect emissions from their business activities. If the locus of influence is beyond the control of the firms' activities, efforts to negotiate and improve their reach would far outweigh the impacts generated. This corresponds to the role of accounting theories in providing a locus of control for environmental impacts, which can reflect the extended domain of organizational responsibilities.

Depending on the literature bridging the (environmental) sustainability and the accounting needs of business firms to define what (un)sustainability and the accounting of it could possibly mean, this might lead accounting theories to pursue multiple directions but without searching for a cornerstone on which diverse views and efforts may converge. Accordingly, we examine 'light' to 'dark green' environmentalism and related developments from the perspective of an ethical approach that differs from the ethical rights theory and instrumentalism perspectives, while working to develop an equivalence wherever feasible (of course, this effort is open for academic deliberations).

Eco-efficiency or the light green view

This is equivalent to the instrumental view, where selective changes are adopted to ensure business gains. The instrumental view of environmental accounting relates to how firms can adapt innovative methodologies to improve their products, processes, conduct and standing in society as forward-looking members. While these aspirations might not force the firms to move away from their current goals (which might not be environmentally superior), the instrumental approach is not concerned or critical about the overall behaviour of the firm but instrumental in letting firms leverage new ideas and improve their current and future goals.

The intent is no different while experimenting with the methodological improvements advanced under EMA or SEA to help firms deal with the need of environmentally sensitive information. This includes waste accounting and reporting methodologies to help firms develop cost drivers, identify environmentally sensitive expenditures that are getting diverted from the value chain and help firms reduce waste as well as control costs. In comparison, MFCA uses mass balance and flow cost accounting to improve the scientific view of waste flows and costs of the non-production outputs that form part of the production cycles. Besides these, firms have embraced advanced LCC and FCA techniques to cover out-of-boundary and difficult-to-quantify external costs that did not form a part of the corporate accounting records.

This shifted the focus to establish the accountability of the firms for externalized liabilities, for which the scholars did not always agree with in breaching the organizational boundary (Jasch, Ayres and Bernaudat 2010; Laurinkevičiūtė, Kinderytė and Stasiškienė 2008; and Lee 2011). However, the advancements acknowledged the need for a mechanism to record the environmental impacts that a firm produces and a diligent way to communicate the impacts to the stakeholders. The proposed advances remained challenging due to limitations in monetizing aspects that remained underdeveloped, and failed to fuel imaginative solutions in any other form that could improve the logic of accounting from the perspectives of polyvocal citizenship, for example, to experiment with a unified construct of a social and environmental management accounting system that can work in coordination with the existing management accounting system, as proposed by Gray and Laughlin (2012). At a minimum, this view helps us to conceive environmental accounting as an instrument to help firms deal with the risks associated with dynamic changes in the legal and accounting world, not to mention responding to societal expectations for firms to be agile and sustainability-oriented.

The environmental or green view

The green view proposes to manage environmental aspects and business externalities in such a way that their temporal fingerprint is minimized or can be turned into a positive and integrated into business strategy. I argue that in the absence of systemic integration of environmental bads, voluntary reporting contributes to information asymmetry, and substantiates it further by invoking the theory of ethical rights, which advocates for the right of society to be informed about how firms might be operating in relation to or contributing to environmental (un)sustainability. Incidentally, this has been a basic premise of the SEA theories as well. However, SEA theories remain less overt about the societal right to be informed about the negative impacts that a business might be contributing to; it largely remained implicit.

For example, a firm's licence to operate is based on its ability to add value to society, but this also needs the firm to be open to scrutiny about the negative impacts that it is contributing to, if any, and the steps it is taking to improve the situation. One-sided information—mostly the positive ones that the CSR and voluntary frameworks disseminate—would limit the field of view of the users of such information and contribute to information asymmetry. Unless the claims of positive action are backed by information on the negatives, it might appear as a deliberate attempt on the part of the firm to substantiate, promote or hide certain agendas that might do more harm than good.

The argument here is that to be reflective of the *real* worth of a firm, both sets of information (ethical positives as well as negatives) are a necessity, including the assessment of whether the current organizational practices of a firm are effective enough in dealing with its impacts or contributing to the overall well-being of society and nature. Examining the impact of businesses on the proximate environment and evaluating negative externalities that are outside the control of the market mechanisms and prevailing economic arrangements of resources (Broughton and Pirard 2011; Dahlman 1979) are direct outcomes of this view, although less prevalent.

Bainbridge (2007) believes that the externalities should be brought inside the accounting boundaries to reflect true environmental costs, which if persistently left outside the economic realm would eventually lead to market failure. Bracci and Maran (2013), while reviewing the accounting reporting of Italy, felt the need for a proactive environmental system that would break this barrier and bring externalities-related costs within accounting and could help firms in integrating negative externalities as part of the social contracting process as well.

Comite (2009) believes that an ethical–environmental balance sheet would reflect the collective relationship of the company where, instead of measuring profits (an economic measure), the question of subsidizing externalities could be raised to define the real profit and sketch future relationships with societal demands and stakeholders. Others have also considered changing paradigms to improve accounting and the control of absolute goods and externalities such that the accounting system could internalize externalities and integrate the impacts as a part of the decision-making system (Hájek, Pulkrab and Hyršlová 2012; Shevchuk 2013).

While the externalized costs can be regarded as the societal load that a firm produces (in the form of environmental assets) that would cost society now or in the future to mitigate or abate, and/or to maintain the constancy of natural capital, there needs to be a mechanism to capture these. Using provisional, social and externalized costs (Lungu et al. 2009), it is theoretically possible to improve the inclusivity of accounting to record costs that are borne by society today or will be borne by society in future to mitigate or abate the impacts that business activities are yet to internalize.

The ecological or deep green view

The deep green view looks to measure and account for ecological losses in terms of the impacts of business activities on biodiversity, inter- and intra-generational equity, and social welfare, corresponding to the ethical approach that advocates firms to view their obligations to nature and society as a primary concern, as an indispensable part of their existence beyond the disclosures in CSR reports. This needs to develop into an uncompromising attitude in businesses to uphold ethical causes as a central tenet of their existence, simply because 'this is the right thing to do' (the normative view). Even though this might not always lead to a win–win situation for firms, especially when the outcome(s) of any strategy is(are) stacked against the economic opportunities that need to be sacrificed to pursue an environmentally oriented agenda, still the expectation for firms would be to base their actions and decisions on the intent to maximize the overall well-being of everyone concerned (Frederiksen and Nielsen 2013). Although it would be impractical to get into the calculative aspects of enacting such an ideology, it is notable that the decision to behave is left to the businesses.

This roughly corresponds to the deep green approach of SEA theories (which advocate caring for nature in everything we do), but is contrary to the methods of economic organizations that have been more in favour of a progressive adoption of environmental sensitivity, that is, the light green approach (Gray and Laughlin 2012). At the same time, operationalization of the ethical approach has remained an unresolved area, which despite the best intentions, is a difficult one, as evidenced in the survey of Australian companies in their use of an EMA toolkit, where the surveyed companies preferred to use newly developed environmental techniques for internal reporting purposes only. The practitioners felt that the current level of theorizing and the lack of practical insights of the new techniques are inadequate for their wider usage (Petcharat and Mula 2012).

Subsequent developments within the SEA theories have somewhat eased the limitations on effective operationalization by using voluntary reporting mechanisms to disseminate information on how they are (or would be) contributing to societal expectations (following from BAO—Belief—Action—Outcome—framework where belief leads to actions), even though the uncertainties regarding the loci and contents of such broadcasts add to the unresolved questions of traceability and assurance of disseminated information that remain disconnected from the formal accounting process (Wilburn and Wilburn 2013) and lend credibility to the hypotheses of a widening gap between the accountability of organizations and the efforts to incorporate ethical considerations as a part of their business conduct (Unerman and O'Dwyer 2007), which are still awaiting refutation. Without negotiating concerns of trust and belief that the firms need to address through disseminated information, a long-term view of these efforts could hardly lead to a meaningful dialogue between the firms and society or to an ensuing debate on the firms' contributions to any form of sustainability (Gray and Laughlin 2012). Moreover, voluntary reporting practices have encouraged the growth of multiple reporting frameworks including the CSR bandwagon, which has witnessed a plurality of standards and languages, where a lack of fact-based or assured information is the order of the day.[1]

The various advances discussed have been arranged by increasing environmental proactiveness in Figure 10.1 to deliberate upon what kind of accounting solution would prove to be a pragmatic one.

SUSTAINABILITY AND ACCOUNTING PARADIGMS

Sustainability-related concerns due to the economic activities of firms and industries correspond to the ecological impacts produced by the firms that firms should account for. Evidently, the deep green view is more all-encompassing than the lighter levels, as it would need the accounting to account for biodiversity-related losses, the permanent loss of natural capital, reduction in inter- and intra-generational equity and the loss of ecological resources. Atkinson (2008) believes that the capital approach could be the way to relate the built environment to sustainability, and thus to evolve a pragmatic solution to evaluate and

---

[1] For example, refer to Albu et al. (2013), Akisik and Gal (2011), Brennan and Merkl-Davies (2014) and Hazelton (2013).

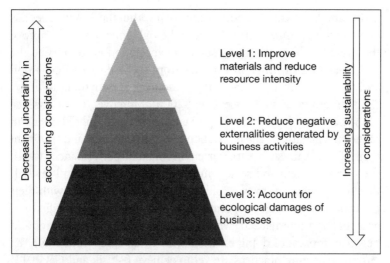

Figure 10.1 *Accounting paradigms of sustainability as part of accounting frameworks*

Source: Author.

account for the impact, for example, by using metrics like the sum of natural and man-made wealth. Jones and Solomon (2013) have contributed to highlight the problems of accounting, accountability and technical complexities regarding accounting for the biodiversity-related losses that corporate firms are responsible for. Metcalf and Benn (2012) have questioned the 'fit for purpose' of corporate firms as part of the social mechanism and highlighted the changing paradigm of the role of corporate firms that needs to be taken into account, as well as the need to consider the interconnections they have with the dynamic economic, environmental and social systems, which in turn goes along with the need to have ecological and/or sustainability accounting standards (Akisik and Gal 2011; Aras and Crowther 2008; Schaltegger 1997). In this regard, Frankham et al. (2012) have used empirical data to highlight that accountants (as practitioners as well as profession) are rarely involved in sustainability-related initiatives and their strategic integration within firms; however, the authors also hoped for an improved scope of accounting services to change the attitude of firms by contributing to areas such as risk measurement, strategic integration and review or assurance activities.

In summary, while the ethical approach within the SEA theories has depended on business firms to act in the righteous way and improve the overall welfare of society, such an expectation is contrary to the real behaviour of firms, if learnings from the Industrial Revolution and the current state of the earth's resources are anything to go by (Millennium Ecosystem Assessment 2005). Moreover, in the absence of the necessary legal framework to enforce firms to uphold their end of the ethical bargain, the motivation to behave ethically and (re)orient their conduct towards superior environmental performance (which the market system and society in general would expect and reward) demands a mechanism that calibrates firms in accordance with their contribution towards this goal and which can easily be relied on, accepted and internalized, while reflecting in detail the positive and negative impacts of doing business with equal ease—something the current reporting practices are yet to achieve (Akisik and Gal 2011; Negash 2012).

Referring to the accounting paradigms of encapsulating sustainability (Figure 10.1) and considering the limitations of the accounting language in handling transactions that use a closed-ended approach, any accounting framework would have to restrict examining the firm–environment exchange to a particular level (say, first-order impacts only) and bring only these within the boundaries of bookkeeping (level 2). Although such a vision might be a curtailed one and not entirely reflective of how sustainability might be\impacted, this would still be a practical approach, and ahead of the conventional eco-efficiency based outlooks or disclosure-based reporting (level 1). It offers a hope that the accounting theories may catch up and achieve complete accounting for biodiversity-related losses (level 3). Further, if such an arrangement can establish the much-needed causal link between the firms and the environment that the accounting mechanisms are known to look out for, it would be a way to enact the accountability of firms for the environmental assets they generate, a solution worth leveraging. Accordingly, this discourse can lead to the development of an accounting mechanism that pursues the middle path to cover the environmental issues that businesses generate, explored further and experimented with in the next chapter.

Environmental Accounting

An Independent Accounting Viewpoint

INTRODUCTION

Continuing from the previous chapter, where the discussion regarding sustainability challenges for firms converged to evolve a suitable representation, this chapter discusses framing a feasible enactment of such a vision. This leads to revisiting the generic accounting process and capture associated stakeholders' expectations. To develop a verifiable account of environmental performance, an accounting framework would need to account for environmental impacts by leveraging the transactional repository of business activities to define *what* needs to be accounted for while a process is needed to define *how* to accomplish this while maintaining the mandatory aspects of accounting and an audit trail. Although this has been provisioned in the literature (discussed in the Chapter 'Environmental Accounting: Connecting Critical and Normative Theories of Accounting'), there has been no significant advancement to formalize a solution. In this chapter, the feasibility of environmental accounting is discussed as a pragmatic solution to tie environmental impacts into the business activities and their temporality to translate these to externalized liabilities, bringing much-needed quantitative aspects to the externalities and helping businesses to be cognizant of the real barriers. The practical aspects of such a framework and its ability to handle real-life cases are explored in this and subsequent chapters.

FRAMING ENVIRONMENTAL CONSIDERATIONS AS A PART OF ACCOUNTING

If we consider the accounting practice as a language to interpret business activities, its enactment needs to be based on the norms and standards that are a part of a framework through which the business transactions are analysed (Merino 1993), where the theoretical underpinning of the framework provides the generalization of 'what accounting is *accounting*'. This leads to methodological challenges in handling firm–environment interactions. Demands placed on an accounting framework to evolve and meet the information needs for firms in a continuously changing business environment can hypothetically be achieved by stretching the current frameworks, one would hope. This has not been found feasible so far, however, as we discovered from the discussions in the previous chapters, simply because:

- The sustainability-related information needs of firms and stakeholders are not within the economic existence of organizations.
- Recent developments cannot be encapsulated by improving existing accounting methods.

Findings from previous chapters and advancements that led to different enactments, explored previously, and supports following conclusions:

- First and foremost, it has been unequivocally proven that planetary well-being is everyone's responsibility (being our common future), including concerns like inter- and intra-generational equity, over-utilization of resources, uneven distribution of material well-being, depleting resource levels, poverty and other aspects. They all relate to sustainability, which is beyond the economic sustenance of firms.
- As the stakeholders' need to engage with the environmental implications of doing business is not the same as the firms' financial performance, this need cannot be satisfied by financial or cost-accounting frameworks. Moreover, it would be a stretch to imagine a seamless integration of these considerations within the basic schema of these frameworks as that is neither the intent of the existing frameworks, nor their lingua franca.

- Arguably, reinforcing the environmental accounting needs within the existing accounting frameworks—not to mention accommodating ecological accounting standards as a part of GAAP—would pressurize the existing frameworks to rewrite the rules, which would be time-consuming and could severely compromise the cost and quality of information currently available. As the IFRS has called for the global standardization of accounting practices, it is not designed to reflect regional, local or environmental diversities or imbalances, not to mention other environmental considerations.
- Recent developments within EMA have already detailed how it has furthered methodological advances for environmentally conscious decision-making and considered externalities to a certain extent, while external reporting through CSR improved the external dissemination of organizational intents and activities regarding social commitments. However, the efforts are still falling short of building an accounting system that is transactional and develop a rational basis to cover the impacts generated by firms that could be standardized as a part of the organizational practice, more so because it involves environmental and social implications of doing business, which brings in concepts such as externalities, social costs and the constancy of natural and other types of capital that are alien to the current accounting standards and practices.

This leads us to conclude that:

- For the firms to handle externalities better, information needs to be based on a uniform interpretation of its business activities without losing temporality, transparency and traceability.
- For an accounting framework to accomplish such an objective, it should also address overlapping accounting principles (formal versus informal types of accounting principles), measurement differences (financial versus non-financial or mixed units) and unstructured information content (nature of externalities).
- Although an accounting framework with similar characteristics has been envisaged as 'environmental accounting' in places, there has been virtually no progress to develop it as an integral part of the organizational accounting function.

- Without commenting on the feasibility of such a construct, the minimum condition to be fulfilled by such an accounting system is to leverage a suitable measurement schemes and use the language of accounting to provide interpretations of firm—environment exchange in a way that connects to the ethical theories.
- The environmental considerations of doing business can be viewed per an entire range of perspectives (from eco-efficiency to a deep green view, per Figure 10.1). A suitable enactment to realize these views would need an encapsulation of the intent that is immediate and bounded, maybe by restricting itself to first-order environmental impacts of business activities.

The points converge to conceive 'environmental accounting' as an independent accounting framework that can encapsulate the accounting considerations of environmental stress in a scientific way. While this leads to the epistemological challenges that a new framework would bring in—for example, in defining how social costs and externalized liabilities impact environmental assets and liabilities, and discussing how this new framework would differ from the way accounting traditionally related to assets and liabilities—it is ultimately challenging for conventional accounting norms to open the doors to contractual obligations that are based on fair and equitable business practices beyond economics. To relate to some of these vexed issues, the scholarship is synthesized and, explored next.

INTRODUCTION TO A DIMENSIONAL VIEW OF THE ACCOUNTING UNIVERSE

If environmental information is to be captured within the accounting framework, and data and information generated by accounting processes are to be considered as symbols, their utility lies in the usefulness of the viewpoint that they provide within the broader social perspectives. In that sense, accounting theories are inherently pragmatic in nature (Cowen 1968). The role of pragmatism in this case is not to dispute the pluralism of meanings, but to verify whether these meanings can be generalized to guide theory and practice and converge even within the changing environment. Accordingly, the relevance of any methodological improvements would depend on how the newfound knowledge translated into improved knowledge of organizational performance. Prevailing environmental accounting methodologies have so far developed only isolated views of the aspects and offered methods that are (mostly) computational in nature, yet less relevant to the accounting process.

Building an accounting view of environmental impacts: the ontological basis

Thomas Kuhn (1996) elaborated in *The Structure of Scientific Revolutions* that growth of knowledge happens through paradigm shifts, where old theories cannot respond to anomalies without stretching their boundaries beyond the existing paradigms, until new questions can no longer be addressed through the existing theories and demand more comprehensive frameworks that subsume the old theories and expand further. Consequently, any form of accounting intervention that intends to help firms with information on the environmental impacts they produce cannot remain confined within the current accounting methodologies as they are institutionalized today. Logically, these impacts would have to be based on the damages that they cause to the proximate environment (the environmental view), if not to the entirety of humanity and to all ecosystems (the ecological view), calling for restoration and redressal. Unless the first level firm-environment interactions are formalized as environmental obligations, firms would not be operating any different than today. Other than allowing the existing frameworks to continue, information generated by using the data from such a construct would have to be auditable, verifiable and to have its own set of rules and standards to follow. Accordingly, an independent accounting framework should be equipped to capture and translate the environmental perspectives of business transactions and encapsulate them within an accounting schema to offer a temporal view of the impacts.

To further the development of an independent accounting framework to handle environmental considerations, I would like to draw upon the origins of cost accounting and reflect on a few valuable insights it offers. Cost accounting evolved as a response to the constraints faced by decision makers in extracting information on the value-creation process from the financial accounting of transactions (Edwards and Newell 1991). The need for better information on the value-creation processes and the flow of costs led to the development of cost accounting and its techniques, which grew (mostly) outside the formal structure of financial accounting. By leveraging the double-entry principle, cost accounting legitimized the accounting of materials and resources and satisfied the management's need for accountability, control and decision making support (Hoskin and Macve, 2000). In due course, cost accounting was established in the accounting literature as a separate and independent accounting framework, one that could satisfy the specific boundaries of accountability around transactions and serve stakeholders' interests (e.g., financial well-being—external stakeholders information versus overhead cost analysis—internal information needs) and the decision-making needs of management. Even though in time, cost accounting formed a part of the integrated system of accounting, its integration had more to do with using the same form of measurement that is being used by the financial accounting system and less to do with the inherent viewpoints that the respective frameworks continue to serve.

In the evolution of the accounting sciences, we are yet again at a crossroads as we experiment with the accounting of environmental impacts. Taking into consideration the inherent limitations of the existing accounting frameworks, a separate and independent viewpoint of environmental accounting is better equipped to help firms bookkeep environmental aspects and impacts (similar to accounting of stock quantities and material values within cost accounting) to generate verifiable bases for information for environmentally conscious (internal) decision-making and (external) reporting requirements. However, challenges to this approach relate to the porous contractual boundaries between the firms and society when it comes to covering environmental damages (other than legally mandatory actions) and the indeterminate state of ownership of waste. The ownership of wastes has been discussed at length in Chapter 5, where the perspective of residual ownership becomes an active ingredient to encapsulate 100 per cent of the liabilities for first-order impacts, beyond which the locus of absolute control fractures. This pushes environmental accounting to cover first-order environmental impacts at a minimum. Again, we must assert our requirement of accountability to ensure that the firms are not working towards economic gains alone and are not exploiting the situation due to missing legal boundaries, as has been the case in developing countries. Granted that this is still far from achieving a complete accounting of biodiversity-related losses, it nevertheless paves a way for accounting theories to catch up to social expectations.

The discussion should not to be construed as the concluding comments on accounting for environment. Environmental accounting as discussed here is limited to the bookkeeping of environmental impacts produced or sequestered by business activities and not meant to fulfil the need to evaluate the ecological sustainability of businesses, which has been discussed in the last chapter. This latter includes the need for a corporate response towards ecological challenges and their accounting (ecological accounting), which is being discussed here to emphasize that an *all-inclusive* accounting framework might be an ideal choice, but its enactment is elusive at this point of time. At this juncture, it would be pertinent to clarify that the words 'accounting viewpoint' or 'accounting dimension' used interchangeably throughout the remaining text reflects the isolated boundaries of common stakeholder considerations and handle the firm–environment exchange, and in no way reflects the notion of dimensionality proposed by Ijiri (1987).

Environmental view of business transactions: Epistemological expansion

In generic terms, by analysing business activities, any accounting framework can abstract accounting information and record the relevant information in the books of accounts. In generalized terms, transaction 'X' in time 't'—representing an organizational event connecting process state 'A' to 'B'—could be analysed by an accounting framework in accordance with the viewpoint that it upholds and the relevant accounting rules applied to create the necessary accounting entries. While financial accounting records the material value of a transaction and the direction of transfer of contents, cost accounting analyses the transaction from the perspective of value creation. On similar lines, a separately established environmental accounting construct should reflect how a business transaction interacts with the environment and abstract the related information to generate corresponding entries in the environmental accounting records. This construct captures corresponding accounting interpretation of firm—environment exchange that would be separate from the financial and other accounting needs or from the need to consider these as a part of the existing frameworks (Figure 11.1).

Figure 11.1 generalizes that accounting frameworks analyse transaction 'X' to create bookkeeping entries in accordance to the viewpoints that the framework holds itself committed to. As a result, environmental accounting could encapsulate the environmental impacts of inter- and intra-firm transactions while remaining independent of the financial and cost accounting entries, yet be connected to these by the same underlying transaction, and could develop a transactional base of such impacts over time.

Environmental analysis of business transactions: methodological improvisation

The next logical question would be what would constitute the environmental view of a given business activity. How do we analyse and abstract what needs to be accounted for? The environmental impact of any transaction is understood at two levels. First, the physical characteristics of transactional exchange, and second, the resulting damage caused. The first part of the exchange is the aspects and the second part of the exchange is its impact. This changes the notion of assets and liabilities from the way they are defined in traditional accounting parlance, due to the attached economic considerations. Here, environmental considerations redefine environmental assets as the physical assets that the firm generates due to its business activities, while impact

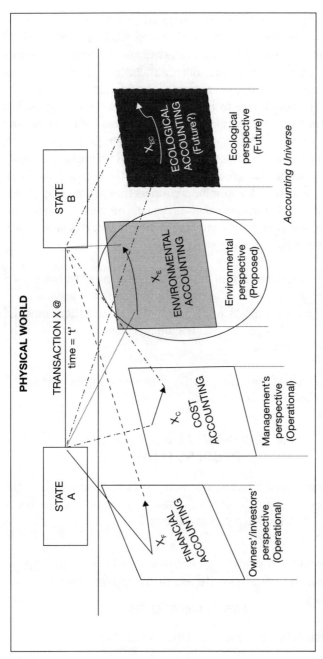

Figure 11.1 *Accounting implications of a single transaction in parallel accounting universes*

Source: Author.

reflects efforts at countering the environmental damages that can be caused by the physical aspects. Moreover, in different circumstances, the aspects may not reflect physical assets but are derived from calculations, whereas impacts may not always have a concrete value but can be an approximation. To respond to the information needs of the firm, aspects would have to be traceable and verifiable and should form part of an overall transactional system, instead of being derived from computational black boxes. Before proposing a design for such a system, first step would be to analyse business transactions to separate the environmental aspects and impacts at the transactional level. Table 11.1 analyses a few common business transactions using three different accounting frameworks—financial, cost and environmental accounting—to generate a comparative view. From this table, a few key points emerge:

- Every business transaction in a firm can be scanned within a given industry or environment to cover the aspects that it generates.
- Environmental aspects need norms to translate them into the impacts that they produce. This should be handled as a part of the accounting process.

Table 11.2 reflects the comparative positioning of environmental accounting and other frameworks.

ENVIRONMENTAL ACCOUNTING PROCESS FLOW

Given the proposed structure of environmental accounting, now the process flow of environmental accounting can be discussed. It would have to follow a four-step process (Figure 11.2).

Systemic integration of environmental accounting within the corporate accounting framework

Step 1: Identification

Select a transaction and ascertain types of aspects involved.

Table 11.1 Comparative accounting viewpoints for sample business transactions

Business transaction	Financial accounting	Cost accounting	Environmental accounting
1. ABC Inc. receives 100 tonnes of lye at Warehouse 1 against PO# 131.	Book liability for the materials received as per the agreed PO rate.	Receipt inward of 100 tonnes of lye in Warehouse 1 and increase: a. Lye stock b. Stock value.	Environmental aspects generated due to emissions from transportation of the materials.
2. Dispatch of 1000 tonnes of coffee grounds from Jakarta to Rotterdam via ship by Coffee Inc.	a. Payment of shipping charges b. Raising of an invoice and a booked sale of 100 tonnes of coffee grounds.	For the dispatched quantity: a. Reduce inventory b. Account cost of goods sold.	Environmental aspects generated due to the shipment of products by sea route.
3. Residence Hotels has completed 6,000 guest nights worth of business.	a. Book revenue for delivered guest nights. b. Adjust advances received against invoices raised.	Book costs due to consumption of: a. Materials b. Resources such as labour, energy and water.	Environmental aspects generated due to: a. Energy use b. Water use c. Waste generated d. Others, if any.
4. 400 hours of consultancy services towards system development by IT Services Ltd.	a. Booking of consulting expenses for 400 hours b. Travel and lodging expenses of consultants.	Not applicable	Environmental aspects due to: a. Energy consumption by software, server and programmes b. Travel undertaken by consultants

(Continued)

Table 11.1 (*Continued*)

Business transaction	Financial accounting	Cost accounting	Environmental accounting
5. Disposal of 100 tonnes of solid waste through contracted services.	Booking of expenses towards 100 tonnes of waste disposed at the contracted rate.	Cost of waste disposal charged to cost centre.	Environmental aspect of waste disposal depending on the disposal option.
6. Salary and wages disbursement to employees.	Salary and wages charged to regular expenses account.	Salaries and wages are recovered against overheads charged.	Not applicable.
7. Facility services availed towards cleaning and housekeeping.	Cost of services charged as accrual expense at month end.	Service cost absorbed as part of general overheads.	Environmental aspects of services like water, energy and emissions are accounted.
8. Jobs subcontracted to service providers.	Subcontracting expenses are booked on receipt of materials.	Cost of subcontracting is absorbed against the product cost.	Environmental aspects produced in subcontracted jobs are inventorized.
9. Funds invited for investments in new schemes.	Collected funds remain in the balance sheet against the scheme till disbursed or utilized.	Not applicable	Not applicable
10. Kyoto units to be purchased from the market due to carbon shortage.	Cost of purchase is expensed against the shortfall.	Not applicable	Purchased Kyoto units are added in the environmental asset records.

Source: Author.

PO = purchase order

Table 11.2 Comparison of environmental accounting with other frameworks

Transaction	Financial accounting	Cost accounting	Environmental accounting
1. Transactional coverage	With external entities (inter-unit)	Intra-unit (internal) transactions	Inter- and intra-unit (all external and internal) transactions
2. Bookkeeping coverage	Accounting ledgers in financial terms	Inventory (quantity-based) and cost ledgers (value-based)	Aspects accounting (quantity-based) and environmental ledgers (value-based)
3. Unit of measurement	Only monetary unit	Physical and monetary units	Physical, monetary and/or mixed units
4. Valuation norms	Historical costs, fair market value	Historical costs	Multiple methods (based on the type of aspect and the organizational policies)
5. Primary target audience	Shareholders and investors	Internal decision makers	External and internal stakeholders
6. Dependency	None	With financial accounting	With financial and cost accounting

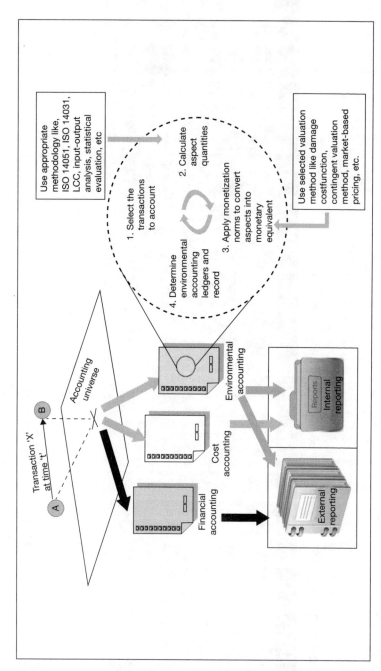

Figure 11.2 *Environmental accounting framework*

Source: Author.

Step 2: Quantification

Decide on the technique to calculate the aspects in the applicable units of measurement (UOMs). The quantification technique can be a standard one and/or approved by the EMS of the organization. The EMS being the owner of and the expert on these environmental aspects, it can recommend suitable techniques such as ISO 14031, ISO 14051, input–output analysis or LCA to establish the corporate parameters for determination and quantification of different types of aspects, rightly absolving the accounting function of the responsibility of identification and quantification of aspects.

Step 3: Monetization

Once quantified, the aspects might need suitable methods of conversion to be translated into the equivalent monetary value. The monetization process is not a scientific cost ascertainment and might follow any of the valuation methods (e.g., at cost, at replacement cost, market determined rates, contingent valuation method, cost avoidance method, or any other combinatorial proxy), so long as it can be used consistently and is approved by the management, reflecting the intent of management to view the externalized liabilities. Any change in the valuation method would generate adjustment entries to reflect the changes in environmental accounting (discussed in detail in the next chapter).

Step 4: Ledgerization

The final step will be the ledgerization of the quantified and/or monetized aspects in the aspect ledgers, using a standardized journal entry process.

The resulting interrelationship of multiple accounting constructs is the natural outcome of using accounting language to interpret transactions according to the domain of interest. The commonality of the business transaction imposes the interconnectivity between the viewpoints. For example, 'receipt of materials' is recorded in cost accounting as a transaction that increases the inventory and generates financial accruals. Any proposed construct would need to check for environmental aspects generated due to the 'receipt of materials'. Since Scope 3 emission (due to transportation of materials) may be associated with the transaction, the emission aspects would be quantified, monetized and recorded in the environmental books. Depending on the number of viewpoints, every transaction would fall in any of the seven (or $2^n - 1$ interactions, where 'n' is the number of dimensions) types of accounting relationship (A to G, in Table 11.3).

- Type A transactions would impact in all dimensions—for example, purchase and sales transactions. These may involve exchange of inventories and financial obligations, and may generate environmental implications.
- Type B transactions would cover materials and financial entries but would not impact the environment—for example, revaluation of on-hand stock quantities.
- Type C transactions would create only financial and environmental impacts. Purchase and trade of Kyoto units from the emissions market would be an example of this type of transaction.
- Type D transactions are of a true financial nature—for example, stock issue, interest on working capital, borrowing money, funds transfers between units, amongst others.
- Type E transactions are a result of production activities and are recorded in the cost and environmental dimensions.

Table 11.3 *Transactional interrelationship within accounting dimensions*

Transaction type	Financial accounting	Cost accounting	Environmental accounting
A	Yes	Yes	Yes
B	Yes	Yes	No
C	Yes	No	Yes
D	Yes	No	No
E	No	Yes	Yes
F	No	Yes	No
G	No	No	Yes

Source: Author.

- Type F transactions would cover cost-specific adjustments and do not create any financial or environmental impact—for example, overhead cost allocation is restricted within cost accounting.
- Type G transactions are of a true environmental nature and without any economic or material impact—for example, changes in the Kyoto units due to a revision in emission calculation.

Schematics for bookkeeping: Environmental aspects generated by business transactions

This section provides a scheme to analyse business transactions from the bookkeeping perspective, reflecting how the corresponding business activities might be generating or subsuming environmental assets. While the transaction categories covered here are not exhaustive and can be enhanced subsequently, the accounting schema represents how environmental accounting separates the environmental dimension of business transactions from their financial and cost accounting.

Transaction type 1: Activities that generate environmental aspects

This category of transactions would result in the generation of aspects like emissions, solid waste, waste water, and so on, that add to the stock of environmental assets. The corresponding liability would reflect an environmental contingency arising due to addition to the common pool, where the accounting treatment would be:

Dr.   Environmental asset (aspect type)   Aspect Qty X – Value
    To Environmental liability (corresponding transaction class)

Transaction type 2: Sequestration or transfer of environmental aspects

Business activities that would result in sequestering or transferring of environmental assets are part of this set—for example, the reuse and recycling of food waste using an in-house vermicomposting facility or recycling of waste water to improve grey water usage to reduce environmental load. Similarly, the sale of electricity by utilities would result in a transferring out of GHG load from the producer to the consumer(s). The journal entry in this case could be:

In case of sequestration:

Dr. Environmental liability   Aspect Qty saved
     sequestered   X—Value
         To Environmental Asset (aspect type)

In case of transfer:

Dr. Environmental liability   Aspect Qty saved
     transferred   X—Value
         To Environmental Asset (aspect type)

Transaction type 3: Business activities earning environmental credits

Involvement of firms in community activities would result in reducing local waste and save social costs—for example, reducing community waste by using organizational facilities, thereby helping the business to earn environmental credits. The journal entry in such cases would be to create a credit (or reward):

Dr. Environmental savings   Aspect Qty X Social costs
     generated   saved
         To Environmental/Social Cost saved

Transaction type 4: Permits, fees, legal charges and other environmental expenses incurred by businesses

These transactions are driven by organizational interactions with the market and the legal system to improve or regulate environmental and social impacts of the firm and would include expenses incurred in purchasing or selling permits or licences, or any other expenditure incurred that is related to or impacted by environmental obligations and decisions. These transactions would generally be accounted within the financial books and lead to environmental accounting also accumulating the financial impacts to enable environmentally impactful decision-making by the firm.

Journal entry for expenditures:

Dr.   Environmental Expenditure   Amount incurred
       (individual head)
           To Environmental contingency covered

Journal entry for income:

Dr.   Environmental contingency impacted   Amount incurred
           To Environmental Income (individual head)

Transaction type 5: Adjustment transactions in environmental ledgers

These entries would take place within the environmental ledgers to transfer balances, enter corrections or revalue aspects due to changes in quantifications or valuation norms for the aspects. The journal entry would be:

Dr.   Environmental Ledger A   Change in value
                To Environmental Ledger B

GENERALIZED CONSTRUCT OF ENVIRONMENTAL ACCOUNTING AND SUSTAINABILITY

This brings us to an exploration of the practical relevance of environmental accounting and its capabilities from the perspectives of information generation and decision-making that characterize management accounting. These characteristics are explored as part of a general framework to validate the relevance of the information for management accounting.

Traceability

Environmental accounting creates the transactional backbone of environmental aspects to be used by the information system to extract relevant and verifiable data for environmentally conscious decision-making. This would absolve EMA and the corporate management information system (MIS) from the need to use arbitrary methods to quantify environmental aspects and would improve the temporality, transparency and traceability of information. Traceability can link a piece of information to the source business event or transaction and to other accounting frameworks.

Timeliness

Environmental accounting opens the door for the organizations to actively consider externalities as a part of their business activities and improve transparency in reporting. Waste creates externalities upon leaving the area of private ownership. Accordingly, the information is available as soon as the activities are recorded and remain tied to time, improving the accountability of the firms.

Relevance

Environmental accounting allows the existing accounting frameworks to continue in an 'as-is' form, which save the time and resources that would be needed to modify these and institutionalize environmental accounting within a single framework. This in turn helps in developing and implementing standards for sustainability and ecological accounting separate from the existing accounting standards. Accordingly, the information would be relevant and possible to trace to the transactions causally.

Uniformity

Environmental accounting separates the computational complexities of quantification and monetization from the accounting process. Needless to say, the identification of environmental aspects and their calculation to generate an environmental input–output inventory would need engagement and active cooperation from environmental experts, while accounting of the aspects through valuation and ledgerization processes can remain within the accounting functions of the business. This calls for the EMS to take an active role in visualizing the environmental activities of a business as cross-functional activities, which would seek active cooperation from other business functions, thereby establishing uniformity of interpretation as well as dissemination of information.

Although SEA evolved as a philosophy to highlight the need for firms to acknowledge accountability for environmental and social externalities and EMA provided the methodological support to influence the internal decision-making processes, these efforts did not challenge the accounting processes to change significantly. However, the prevailing methodologies are now struggling to help organizations go beyond the obvious and failing to integrate the 'external' costs of waste mitigation that society would have to incur at present or in the future. The inadequate coverage of the under-defined ownership of waste and the inheritance of economic theories in accounting has contributed to shape a corporate accounting system that ignores externalities.

Stakeholders' concern over the environmental performance of organizations might create a tension by challenging market-oriented performance reporting, so that the firms would need capabilities to handle environmental performance reports alongside the financial or economic ones. At the same time, it would be a futile exercise if environmental proactiveness were to depend on the results of accounting, it could be highly relevant for firms who have adopted some form of EMS. They would need information that can be developed through environmental accounting, bringing quantitative aspects into the purview of organizational efforts, including setting hard targets and measuring organizational performance against these (explored in Chapter 17). Moreover, the feasibility of environmental accounting collaborating with other environmental methodologies (emission accounting, life-cycle methodologies, flow cost accounting and GHG accounting) and other accounting frameworks is something that firms should be able rely on and develop a flow of environmental information for. Well-defined information is critical for businesses to take decisions that could reduce negative impacts on society and the planet and improve ecological contributions towards sustainability.

Environmental accounting discussed in here offers following key takeaways:

- First, the framework is relevant in explaining how accounting has evolved from a two-dimensional construct of time and money into a language to decipher the business transactions. Environmental accounting extends it into the third dimension by measuring the impacts of business transactions on the environment and resources. This encapsulates separately identifiable stakeholders' demands and organizational interests to identify the environmental impacts of businesses.
- Secondly, environmental aspects of human activities and their impacts on the biosphere—in the form of organizational interactions with nature—are subject to the cognitive limitations of human knowledge and would remain so limited until the intricate nature of the cause-and-effect relationship between human activities and ecological responses over time are well understood. These cognitive limitations require us to explore the overall cycle of natural interactions. This supports the layered nature of externalities and limits its 'complete' view. Accordingly, all impacts of an aspect would not be known always, and so our efforts to derive the costs (of abatement or opportunity costs) to abate or harvest the aspects would be limited, qualifying sustainability as an ongoing effort included in instituting ethical considerations for businesses.
- Further, there is no well-established knowledge base for the environmental aspects of businesses and their possible monetization, and this will remain a work in progress in times to come. Instead of developing a solution that is dependent on the frozen costs of each element, environmental accounting is in-principle accounting framework and a placeholder for all aspects—present and future, monetized or otherwise—mirroring the ability of financial accounting to cover all types of business models and their transactions.
- The cost of externalities is not an absolute. It depends on the path chosen after disposal of the waste and on the average cost of disposal set by the society. In an altruistic sense, it is the externalities generated by the chosen path of waste disposal that should reflect in the externalized costs of an organization. Although the generation

of environmental aspects is a natural outcome of a process (in the form of waste), the impacts are dependent upon how the disposal process is impacting the ecosystem. Ideally, waste needs to be channelized in a way that leads to no impact or better if it can contribute to a positive externality
- As the causality linking waste to its impacts is diluted with an increase in distance from the origin or centre of the activity, it would be meaningful to use monetization as a method that is not dependent on the causality of the events but can be adapted based on standardized conventions. As the discarded waste loses identity once it enters the common pool, its environmental impacts become increasingly generic. Considering the general loss of causality beyond the first impact zone, a standard convention improves equalized treatment of how waste impacts are seen. However, where waste can also maintain causal link to the source—for example, radioactive wastes or environmental accidents (oil spills, gas leaks and other isolatable incidents)—the identified impacts are the first-level ones and accordingly the cost to mitigate the impacts can ignore time-delayed impacts (more of this in the next chapter).

In this chapter we explored how environmental accounting framework supports the overall accounting process and businesses by creating the transactional backbone of environmental interactions that could be used by the organizational information system to extract relevant and verified data for environmentally conscious decision-making. This absolves EMA and other management information systems from the need of using arbitrary methods to quantify and derive environmental information, opening the door for the organizations to actively consider externalized liabilities as a part of its operational spillage and making itself 'accountable' to these. All of this would go in-vain unless environmental accounting framework can resolve the monetization challenge, and translate monetization into a computational complexity, delineating it from the accounting process—we are poised to explore in the next chapter—and expand the boundaries of accounting by absolving it from remaining subservient to economic interests of firms, thereby reducing its dependency on monetary unit as the only numéraire.

Advancements in Costing Models to Handle Externalities

12

INTRODUCTION

Before exploring how an environmental accounting framework can be tested and how insights gained from certain case studies may be used, this chapter details the constructs needed to model the monetization of externalities. The modelling exercise contributes to build how externalities can be evaluated as social liabilities while appreciating limitations (explored previously) that have been a lexicon in environmental accounting theories. However, all that the valuation models achieve is to bring a certain degree of objectivity to the scientific and calculative aspects of the evaluation process. Building costing formulations and related exercises is expected to help businesses appreciate the complexities as compared to capturing the impacts of waste in nominal terms. Any valuation exercise would accordingly depend on a common unit of measurement that we can agree to, following which a numéraire can be applied. However, the efforts are complex and limited where the units of measurement are unavailable. One of the common units proposed in ecological modelling to model human–environment interaction is ecosystem service (as a unit), that we can use to derive quantify ecological service units gained or lost due to a human activity. Unfortunately, this is easier said than done, as ecological modelling is yet to develop a common unit of ecosystem services (Atkinson, Bateman and Mourato 2012; Pirard 2012). We are still far from using ecosystem services comprehensively as a standardized unit of measurement. Another area of concern is the complexity of valuation systems, which can range from using various economic proxies including abatement, willingness to pay and restoration cost models to using any specific one. Evidently, the standardization of valuation models is a different discussion and can be a part of the policy economics to be recommended for a specific type of intervention in a region. In this chapter, individual cases are detailed to explore how different valuation exercises can be used by an environmental accounting framework to translate the impacts (monetary) of a specific aspect (physical) or family of aspects.

FULL-COST ACCOUNTING (FCA)

FCA and its relevance in the environmental arena have been covered in brief in Chapter 8. If we refer to previous instances of industrial experiments with FCA within the perimeter of the environmental costs of industrial activities, we can see that FCA has produced limited results. Debnath and Bose (2014) have successfully used it to evolve the costs of a municipal solid-waste (MSW) management system, and so others[1] have used FCA to cover multiple segments of costs (such as construction and demolition waste, waste from hotels and restaurants, household wastes, and so forth). Reference of FCA in the case referred was the cost of solid-waste disposal; beyond the accounting books of the municipality (or the industrial unit performing the operations of MSW management) and included the environmental and social costs associated with the resources. A similar approach can be used by corporate firms to align their computational needs by relying on FCA, instead of investing time and effort to developing an entire gamut of computational models to capture every single perceivable cost element. The case is equally relevant to our impending discussion as it involves using different proxies to develop a cost function; it can also be used along with economic cost approaches such as damage function, willingness-to-pay and restoration costs to compute costs that cannot be assessed directly and/or might prove difficult to compute. In

---

[1] Readers interested in exploring this further can refer to the cited publication, as well as articles in environment related scholarly publications.

that sense, FCA allows the assimilation of different costing techniques to evolve a cost function, which is tested here using a specific case of evaluating the cost of solid-waste disposal for municipalities.

FCA framework to evaluate cost of municipal solid-waste disposal

Since firms from some industries are dependent on the municipal services of the region to handle a part of their waste, it becomes kind of imperative to assess the social costs of a public system that is external to the organizational boundaries, even when firms are paying for the municipal services. This is true of both developing and developed nations. However, the extent of the externalized costs is dependent on the extent of the marketing forces operating within a region and where businesses might be paying fees for the disposal of waste. While developing countries would have internalized a larger share of the social costs as municipalities would be charging fees to appropriate a portion of their total operational costs, in developed countries, the firms might believe that they are paying for the entire disposal process. In both the cases, what the businesses are anyways not paying for is the time-dilated effects of the disposed waste that are neither estimated nor recovered from the entities contributing to waste generation.

MSW activities and its environmental impacts

Collection

MSW collection entities run manual and mechanical operations to collect garbage, using collection vehicles such as dumpers and compactors to collect waste and transfer refuse from the collection points to sorting or transfer stations. In addition, motorized vehicles are deployed for the placement, transfer and shifting of bins across different locations. All motorized operations to lift, compact, transfer and transport waste need energy to execute, and they generate emissions that add to the environmental load.

Transportation

Compactors, trucks, dumpsters and other motorized equipment transport huge quantities of garbage from collection centres to sorting centres, and subsequently transfer sorted trash to its ultimate destination. These activities consume fuel, which adds to the environmental load due to emission generation. Beyond the efficiency and load-bearing characteristics of these machines, the operational efficiency of MSW transportation is seldom optimized, especially the environmental load and mileage efficiencies. Sophisticated transportation planning systems are available and deployed in for-profit organizations; similar deployment within municipalities would improve operational efficiencies and reduce the fuel consumption of the fleets through optimized route planning, which could also reduce environmental loads.

End-of-life treatments

The end-of-life treatment of garbage involves transferring waste to its ultimate destination. Multiple end-of-life treatment options are available to choose from like recycling and waste-to-energy (WTE) plants for incineration, composting, biogas generation and anaerobic digestion (Saini et al. 2012). Landfilling in such cases is the final disposal method for waste that cannot be recycled and includes waste generated from WTE operations. However, each of these methods would have its own environmental load—in terms of energy use, harmful emissions (oxides of carbon and nitrogen, particulate matter, dioxins, etc.), generation of ash sludge and GHG contributions—and would depend on the characteristics of the refuse being disposed of. On the other hand, informal and backyard recycling in developing economies—for example, of discarded electronic appliances and gadgets—remains outside the market forces, where manual labour work under unhygienic conditions adds to the social and environmental externalities (Sepúlveda et al. 2010). A better-planned waste cycle would often work towards reducing the overall environmental load of the waste life cycle, instead of targeting the reduction of aspects in individual areas, like reducing leachate discharge or landfill gas (LFG) from landfilling.

Landfilling

Landfilled waste continues to degrade and impacts the environment and natural resources due to further decomposition, or due to the ability of certain waste to remain inert (including non-biodegradables like glass) or due to their tendency to contaminate the soil, underground waste reservoirs and proximate water bodies through toxic leachates or to release landfill gases (LFG) like methane, which would add to the GHG load. Also, a broken recycle chain impedes the return of recyclables to the materials cycle (Sharholy et al. 2008) and adds to the landfilling load, reducing its life and contributing to the avoidable loss of recyclables.

In general, MSW practices are hardly designed to capture and evaluate environmental efficiency and concentrate mostly on meeting the targets set by the civic administration, where financial criteria and resource availability are the primary drivers. Accordingly, efforts are needed to research and incorporate advanced sustainability-oriented strategies that would improve closed-loop material recycling, reduce emissions into air and water, and improve social participation to reduce waste generation as part of policy and legislative goals. Figure 12.1 details the energy needs and environmental impacts of MSW flow.

However, it would also need efforts to institutionalize tools and techniques that are targeted to improve not only the economics and budget considerations, but also reduce energy consumption, acidification, eutrophication and ecological toxicity of waste as part of the disposal activities. Municipalities need to improve their information systems to capture the current levels of such impacts and run what-if scenarios as a part of policy implementations to improvise further (by analysing data that could be part of Table 12.1).

Accordingly, the cost of disposal of solid waste by a municipality is developed using an FCA framework where the total cost of MSW services is the sum of the internalized costs plus the cost of externalities included to the chain.

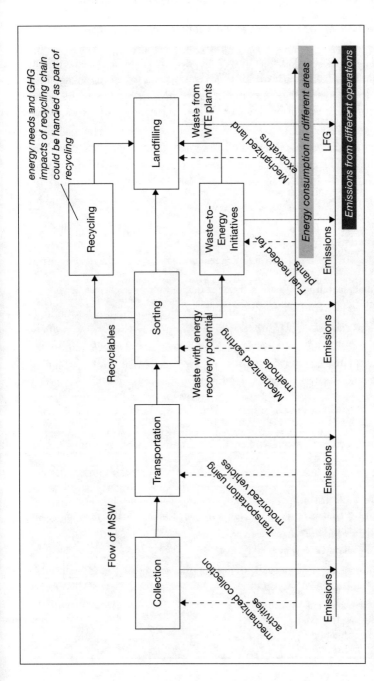

Figure 12.1 *Energy consumption and GHG emissions generated as part of MSWM*

Source: Author.

Table 12.1 MSWM activities and environmental contributions

Activities and corresponding environmental impacts	Collection	Transportation	Segregation	Landfilling
A. Transportation of waste and contribution of motorized vehicles				
1. Compactor	Yes	Yes		Yes
2. Dumper	Yes	Yes		Yes
3. Trucks	Yes	Yes		Yes
4. Tractors		Yes		Yes
5. Refuse collector	Yes	Yes		
6. Large containers			Yes	Yes
7. ...				
8. ...				
I. Total fuel burned = Distance travelled (km or miles) × Fuel efficiency of vehicles (litres or gallons per km or mile)				
II. Equivalent GHG generated (tCO_2e)				
B. Waste sorting and treatment				
1. Transfer station			Yes	
2. Compost plant				Yes
3. Engineered landfill				Yes
4. Incineration plant				
5. ...				
6. ...				
III. Energy consumed = Total energy consumed (units)/Number of hours of operation (hours)				
IV. Equivalent GHG generated (tCO_2e)				
C. Information and other systems that consume energy				
1. solid waste management (SWM) information system	Yes	Yes	Yes	Yes

Advancements in Costing Models to Handle Externalities 215

Activities and corresponding environmental impacts	Collection	Transportation	Segregation	Landfilling
2. Other systems				
3. ...				
V. Energy consumed = Total uptime (hours)/Electricity (units) consumed				
VI. Equivalent GHG generated (tCO_2e)				
D. End-of-life treatment and environment impacts				
1. LFG from landfilling				Yes
2. Emissions from incineration				
3. Emissions from other MSW end-of-life treatments				
4. Others				
IV. LFG generated = LFG potential of waste (tCO2e) × Quantity of waste (kg or mt) × Exposure (hours)				
E. Total energy consumed (I + III + V)				
F. Total emission of MSW activities (II + IV)				
G. Total emission from entire MSW waste chain (F + VI)				
H. Total emission per tonne per day of garbage load (g/tonne of garbage handled/day)				

Source: Author.
Note: tCO2e = tonnes of CO_2 equivalent.

Accordingly, the cost of MSW services is designated as SWC at a specific efficiency level

$$= [\text{Cost of collection} + \text{Cost of transportation}$$
$$+ \text{Cost of segregation} + \text{Cost of disposal}]$$
$$- \text{Revenue generated by the municipalities} \qquad (12.1)$$

where all the above costs are the period costs of social solid waste generation mechanisms and can be derived from the books of accounts of the operating unit.

However, revenue or earnings from MSW services

$$= \sum \text{Fees or levies charged to units} \times \text{number of units}$$
$$+ \sum \text{Revenue from sale of valuables from disposed waste} \qquad (12.2)$$

Accordingly, unit cost of waste disposal per TPD (tonnes per day) (in functional currency)

$$= \frac{\text{Net cost of SWM disposal services (in functional currency)}}{\text{Total quantity of waste collected (TPD)}} \qquad (12.3)$$

In general, equation 12.2 should be good enough to cover the social costs of infrastructure, reduced by the fees or charges paid by the firms to use the SWM services. However, environmental hazards that MSW-related activities generate, to be considered by integrating these within the cost function, are missing from the equation. Two possible externalities are incorporated in this construct as an ideation, but they can be contextual and can be refined further or added to redefine the full cost of a chain:

(i) Net savings to the municipalities due to the involvement of the informal sector

$$= \text{Savings in costs due to reduced MSW tonnage}$$
$$+ \text{savings due to extension of life of landfill capacity}$$
$$- \text{economic contributions of informal recycling} \qquad (12.4)$$

where savings due to the extended life of the landfill

= Saving in landfill capacity converted into period of equivalent present value of rental earning of the land (12.5)

(ii) Cost of methane emissions from landfill

= Methane generation potential derived from the degradability of the solid waste × gross calorific value (GCV) of methane × Density of methane (12.6)

This improves the equation 14.1 as:

FCA of MSW services (SWC) at specific efficiency level

= [Cost of collection + Cost of transportation + Cost of segregation + Cost of disposal] + Externalized liabilities due to social and/or environmental impacts − Revenue generated by the municipalities (12.7)

Accordingly, a generalized FCA construct would be:

$FCA_{(a \to b)}$ of activity chain (linking point A and B)

= Internal costs incurred (direct) + Overheads or indirectly allocable costs + Social cost (incurred by a third party and not paid for) + equivalent nominal value of environmental impacts using replacement cost model − Recoveries (if any) (12.8)

A significant part of this definition is the boundary that is set not from the legal perspective of the firm, but the chain of activities for which the FCA exercise is to be carried out. Here, the use of the FCA framework improves the social cost of mitigating *selected* externalities and our collective understanding of the impacts that the aspects produce. Research leads adding a few (not all) of these externalities can significantly alter unit cost of disposed waste, increasing it to as high as 40% in a specific case, which otherwise remains as unrecognized social and environmental overheads, and shifts from the domain of private profits to social costs (Debnath and Bose, 2014).

LEVERAGING THE TOTAL COST APPROACH (TCA) TO BUILD THE COST OF EXTERNALITIES

This subsection builds upon the experiments on costing methodologies to infer that the cost models are usually based on a causal relationship of the constituents and are generally expected to follow a predetermined pattern. However, these methodologies would suffer where the causal relationships are not firm and where the disposal and environmental aspects at any point could follow multiple pathways upon dissemination, resulting in a chain reaction of impacts in the ecosphere but with different severity levels. Translating the problem to the domain of sustainability, if the movement of aspects is such that the constituents cannot always be traced back to the origin once they have reached the common pool, the cost to eliminate or neutralize these impacts cannot be derived directly or be assigned to a single source. In such cases, modelling of the costing relationships would have to depend on other fields of study, some of which can also be based on the outcomes generated by the constituents. Accordingly, the modelling exercise can leverage economic studies, ecological modelling, demographic studies and more, and develop the cost relationships. To develop a generalized cost model in such cases, the ecological impacts of waste and their movement would have to be considered over different biological and ecological receptors through successive layers of dispersion, which would need to include the environmental impacts of an aspect on air, soil, humans and the ecosystem. To explore the impact on a receptor, the flow of the aspects would have to be traced along every branch of dispersion till the desired node is reached (Figure 12.2). Based on the flow of the aspects in the environment, the corresponding impacts can be modelled using a classification tree (to accept qualitative variable—presence/absence or yes/no) and a regression tree if the impacts can be evaluated as equivalent risk numbers (quantitative evaluation). Accordingly, the model would explore the aspects and multiple pathways by branching along the path of diffusion through interactions with multiple entities (Sorvari and Seppälä 2010) as displayed (Figure 12.2).

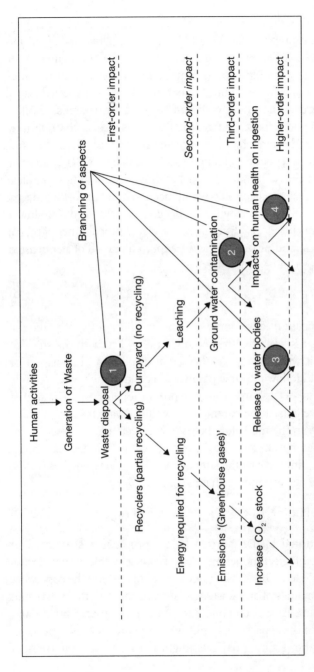

Figure 12.2 *Flow of aspects and impact tree*

Source: Author.

This brings in the need to use different methodologies to aggregate impacts. For example, MCDA could be used to evaluate the aggregate cost of remediation, which in turn is based on the principle of disaggregating a complex problem into a set of decisions or decision theories and modelling the uncertainty by following an expected utility rule to ascertain the underlying course of action (Barzilai, 2010). Similar to the objectives in a tree, the flow of aspects is characterized as essential, understandable, operational, non-redundant, concise and preferentially independent (Franco and Montibeller 2009). Assuming rationality in decision-making and building a set of coherent rules, each intermediate decision step would need to build probability-laced expected pay-offs. Accordingly, if a decision would need to follow a series of events like E_1, E_2, E_3,... with associated probabilities of p_1, p_2, p_3... as corresponding outcomes, the expected pay-off of an outcome 'x' after i steps can be generalized as:

$$E(x) = \sum E_i \times p_i \qquad \ldots \text{for all } i\text{'s} \qquad (12.9)$$

For the purposes of developing costs at an end point i, we take the sum of costs associated with each node multiplied with the share of the burden it would carry. As the actual dispersion of an aspect or element could follow multiple pathways, a child branch will bear only a specific share of the burden and the proportionate costs. The sum of the shared weights at the parent node would always be equal to one. This linearity in the modelling helps to avoid circular references and overloading along a branch. So the overall cost function for an impact 'x' at an endpoint i will be the sum of costs along the branches C_1, C_2, C_3 .... with a corresponding share s_1, s_2, s_3...:

$$P(x) = \sum C_i \times s_i \qquad \ldots \text{for all } i\text{'s} \qquad (12.10)$$

In the traditional MCDA approach, the problem is structured to handle the ecological modelling of alternatives with optimization as the objective. Here, the decision tree is used to develop the cost structure of externalities of any waste as it spreads through the ecosystem. However, there have been concerns in accepting the additivity and multiplicity of the expected pay-offs when the preferences of decision makers and the resulting pay-offs might not adopt an uniformity of expression. Barzilai (2010) is clear in explaining the homogenous nature of the preferences needed to develop a good model. However, such exercises are time-consuming and would need large data sets to run multiple simulation cycles before accepting a few sets. The methodology has been further explored in developing the life-cycle costs of fly ash (FA) that reflects the layered nature of externalities (Appendix), and incidentally, implicit limitations in existing methodologies to relate to the complex web of ecological reactions and losses.

COST VERSUS VALUATION OF ASPECTS: COGNITIVE LIMITATIONS

This subsection contributes to the longstanding debate on developing a costing formulation to evaluate the environmental aspects of waste and reflect its relevance within the environmental accounting framework to provide the 'value' that could be considered as a measure towards monetizing the impacts of externalized liabilities. Organizations might incur certain costs before handing over waste to the societal disposal mechanisms as the first one or two layers of the costs. After that, costs incurred by the disposal mechanism would include the costs of disposal, recycling and reuse, which might not always be paid for by the market mechanisms, and are sometimes part of informal waste recycling as well. Accordingly, the total generalized cost of waste would be:

Generalized cost of waste

$$= \text{Cost incurred by firms} + \text{Cost incurred by disposal mechanism} - \text{Costs already internalized (fees and fines)} + \text{Hidden costs} \qquad (12.11)$$

However, waste could follow different routes of recycling, reuse and disposal. For example:

Each route has its own set of costs and would generate externalities accordingly. In other words, even though organizations can scan the upstream and downstream supply chains and develop an inventory of externalities, the discovery of layers in the chain would depend on accumulating knowledge on all possible methods of waste movement,

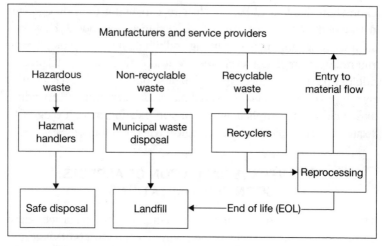

Figure 12.3 *Waste flow through material (re)cycle*
Source: Author.

which is time- and resource-intensive (because investigating the complete waste cycle and its impacts). Moreover, in a dynamic system, this knowledge would always be relative or transitory and not an absolute, given the dynamic state of our collective understanding and the discovery of new ways to treat waste. For example, emissions are the only type of waste that has had a standard route of disposal (at least till recently), which is to dispose of emissions into the air, but our knowledge of its impact on the biosphere is incomplete beyond a point—for example, we do not know how carbon dioxide build-up would interact with the ecosystem beyond a threshold percentage. The problem is compounded by the loss of causal relations once the waste enters the arena of public goods, not to mention the regional complexities and their interactions after dissemination. So, it can be inferred that the quantification and thus monetization of environmental aspects would depend on our contemporary ecological profile of waste, which is *not* comprehensive.

This builds cognitive limitations into a dynamic system, where every single interaction changes the nature of the recipient itself and where the observer is also part of the same system. The limitation in not knowing these variables completely would be due to reasons like uncertainty in ascertaining the exact causal linkages connecting aspects and impacts, including the absence of a single unit of interaction to interpret the damages in terms of losses (cognitive limitations) and deriving costs and the future impact of health and ecological effects, whereas the monetary assessment of damage, remediation and restoration are yet to be established (methodological limitations), and the second- and higher-order impacts of environmental aspects that might not establish causation (interpretive challenges). Thus, every single impact of every single aspect in relation to the ecological receptors and intermediaries can never be known. The same analogy follows for costs that would have to be incurred by society, now and in the future, to handle externalities (Figure 12.4). In other words, there will always be uncertainty in the simultaneous determination of costs (Δc) and impacts of constituents (Δi) within a complex adaptive system beyond the threshold knowledge (analogous to Heisenberg's uncertainty principle, following the deterministic nature of the costs and impacts).

This leads us to two conclusions. First, we would remain information-constrained on the costs within a real system, and secondly, we would have to move away from deterministic rules to accept the probabilistic nature of the relationships involved. In the absence of any standardization to assess environmental damages, different valuation methods have been proposed in the literature to evaluate the costs of waste and emissions. As a result, any construct of environmental accounting would need to develop the capability to work with the range of possible measurement systems, including where numerical equivalents cannot be ascertained with reasonable accuracy. This posits such an accounting system to be different than the traditional form of accounting we have known so far.

ENVIRONMENTAL ACCOUNTING: COST OF EXTERNALITIES AND ACCOUNTING OF ASPECTS

To expand on the theme of the existing accounting framework, it is essential that the prevailing accounting and management information

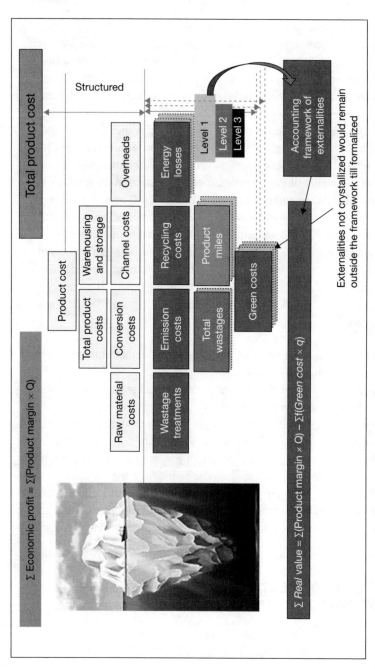

Figure 12.4 *Layered nature of environmental impacts and costs*

Note: Q – Units sold.
q – Units produced.

frameworks are designed to cover information needs on economic health, performance and longevity of firms. These frameworks are designed to fit these needs better than evaluating the ecological preservation of resources. If environmental performance and sustainability-related information needs emerge as the need of the hour, the existing information structure might have to stretch to accommodate these demands. As compared to the economic viewpoints, environmental considerations and ecological preservation of resources are evolving constructs, where the existing frameworks and theories have developed some degree of convergence towards the new requirements by expanding on existing methods, but not beyond what they could assimilate naturally without compromising their original intent. Otherwise, they might compromise their own positions and fail to sustain the existing practices. At the same time, firms would need some freedom to experiment with the available choices.

This chapter has experimented with a few models to develop the costs of difficult-to-quantify externalities by using FCA and TCA. The valuation of externalities has been explored to emphasize the need for new thought processes in costing and management accounting theories in order to incorporate out-of-boundary costs. Since advances in ecological scholarship are yet to establish a standardized unit of ecological interaction, externalities that an aspect would or could generate remain subjective to the cognitive limits. On the same lines, knowledge of how an aspect will behave and interact in a complex adaptive system and the difficulty in isolating these impacts, once the aspects have lost connection with the source (second- and high-order impacts), would remain. Based on these boundaries, it can be inferred that it would be impossible to determine the impact and cost of any aspect with complete accuracy, which translates to a knowledge threshold. This problem needs experts from different fields to come together to support the policy makers in developing and establishing regional averages on impacts and benchmark costs of remediation or avoidance, as the case may be. An environmental accounting framework should be able to handle these concerns. In the next two chapters, some of these considerations are explored through real-life case studies.

Environmental Accounting: Part I

INTRODUCTION

This chapter and the next explore how firms can use environmental accounting to develop a repository of environmental aspects and keep track of how firms are faring in terms of managing the impacts, including objectively working through the actions being taken to improve. This chapter first explains how business activities generate environmental aspects and how these can be captured and/or converted into impacts by working through some examples, before explaining the environmental accounting process with the help of case studies. The gaps discovered and/or confirmed through these cases establish a contextual background for exploring the practical limitations of EMA methodologies. In addition, the studies reflect on the environmental practices of the studied units and the role of environmental considerations in corporate responsiveness. The cases also generated industry-specific learnings that have not always been captured in literature. However, the emphasis of the cases has been less on the contemporary accounting and costing methodologies and more on developing arguments to expand the knowledge base of environmental accounting theories. These gaps are meaningful in establishing why a change is needed in the focus of contemporary research to advance the subject and to develop new insights.

CASE DETAILS

The first case study was conducted in a manufacturing unit and included a detailed data analysis that applied the contemporary EMA methodologies. For more details, readers can refer to Debnath (2016).

Here a summarized version of the case is presented along with a review of the project site, manufacturing environment and operational details, and an analysis of the key findings. Some parts are a repeat from the study report but are included to offer a complete picture.

Project site

This case covers a manufacturing unit (FPP hereafter) for pulps and concentrates of tropical fruits. This unit was a production unit certified per ISO 9001:2004 and HACCP (hazard analysis and critical control points). The study included on-site visits to cover all business processes. Questions related to the seasonality of the industry, the processes and data recording procedures were covered through unstructured interviews conducted with the departmental contacts. Subsequently, operational data were gathered throughout the year at monthly intervals. The unit exclusively processed mangoes from May through July (called the mango processing season) due to the seasonal nature of the fruit, followed by banana, guava, papaya and pomegranate (although not in the same order).

The manufacturing environment

FPP specialized in manufacturing pulps and concentrates of tropical fruits as natural extracts, hermetically packed and sealed in aseptic bags of standard sizes of 1 kg (as sample bags), 20 kg (in corrugated boxes) and 200–250 kg (in steel drums). The products were of standard specifications and sold in local and international markets. The fully automatic, imported plant was manned by a permanent staff of around 40 employees. Based on the production schedules, contract labour was hired during the processing seasons for manufacturing activities. On full load, the plant would operate in two to three shifts of eight hours each. The outline of the manufacturing environment (gate-to-gate cycle) is detailed in the block diagram (Figure 13.1).

Inputs to the process boxes show stage-wise consumption of the physical quantities of materials, water and energy, while the discards are shown as outflows from the processing blocks going as waste to

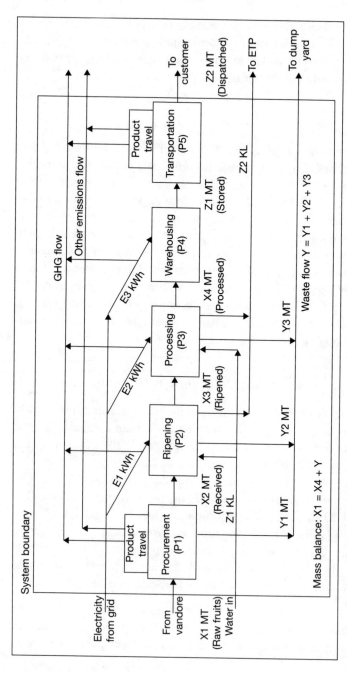

Figure 13.1 *Operational layout of FPP*

Source: Author.

the dump yard (materials) or to the effluent treatment plant (ETP) for waste water. FPP manufactured fruit pulps and concentrates through the sequence of physical operations labelled in the block diagram as procurement (P1), unloading and ripening (P2), processing (P3), and packing and warehousing (P4). The finished products are used by industrial manufacturers as raw ingredients in a variety of food items such as fruit drinks, ice creams, chocolates and flavouring agents. Based on customer orders and dispatch schedules, finished goods were transported (P5) to the customers.

Analysis of key findings

The solid waste, waste water and yield percentage (ratio of pulp output over input quantity of raw fruits) were tracked by FPP as a part of the existing MIS. However, EMA could only add information on the cost of waste and help the management by providing a sophisticated method to redistribute costs, generally borne by the finished goods. The externalities generated remained outside the boundary of the organization. Accordingly, the monetary value of the waste only consisted of the cost of raw materials and resources incurred up to its entry into the waste flow. The stage-wise cumulative production of waste from the case study is detailed in Table 13.1.

A review of the data shows that FPP averaged an yield of 65 per cent, although individual fruit yields were different. In monetary terms, the yield reflected a loss of 33 per cent of the value through the waste stream. The waste disposal process at FPP was limited to the transfer of the waste to the MSW system, where the costs incurred in collecting the waste from the plant and transferring it to the municipal collection point was internalized in form of a payment to the contractors. This cost did not cover the ones incurred by the municipality to dispose of the garbage from the dump to the landfill site and remained out of consideration. A part of the solid waste was also used by the contractor for plantation activities, but the exact usage and quantities could not be verified. Moreover, FPP did not charge the contractors for this diversion. So far as waste water is concerned, it was passed

Table 13.1 Summarized flow of costs on account of waste in FPP

Particulars	Units[1]	Cumulative quantity	Average rate (₹/unit)	Value (₹/unit)
Stage I: Ripening (P2)				
Input (at unloading point)	tonnes	4,650	28,540	132,712,750
Electricity consumption for ripening	kWh	93,895	5	469,475
Total			28,545	133,182,225
Less: Ripening loss	tonnes	492	28,545	14,044,325
Ready for production (to stage II)	tonnes	4,158	28,545	118,691,675
Waste at stage I (to stage IV)	tonnes	492	28,545	14,044,325
Stage II: Processing (P3)				
Fruits issued to production	tonnes	4,158	28,545	118,691,675
Production costs:				
Electricity consumption (for processing)	kWh	570,175	5	2,850,875
Furnace oil (for boiler)	kl	24.7	43,000	1,062,100
Briquette consumption (for boiler)	tonnes	464.86	2,800	1,301,608
Water	kl	24,056	50	1,202,800
Consumption of auxiliary materials	₹/tonne		350	1,455,300
Labour	₹/tonnes		200	831,600
Theoretical finished goods	eMT	4,158	30,639	127,395,958

Less: Production waste (to stage IV)	tonnes	1,151	30,639	35,265,211
Finished goods to packing (Stage III)	eMT	3,007	30,639	92,130,747
Stage III: Packing (P3)				
Packing costs	₹/eMT	3,007	4,000	12,028,000
Support costs	₹/tonne		300	902,100
Overall cost of production			34,939	105,060,847
Stage IV: Waste treatment (P2–P4)				
a) Waste water	kl	21,117	50	1,055,850
Electricity consumptions (for ETP)	kWh	43,166	5	215,830
Cost of ETP-treated water	kl	21,117	60	1,271,680
b) Transport charges for waste disposal	tonnes	1,643	60	98,580
Waste stream from stages I and II	tonnes	1,643	30,012	49,309,536
Cumulative waste stream value	tonnes (solid)	1,643		49,309,536
	kl (water)	21,117		
Fruit yield (%) (on weight basis)				65%
Waste stream (% of total cost) (on value basis)				32.5%

Source: Author.

eMT equivalent metric tons of inputs (raw fruits)
FG *Finished goods*
[1] KEY

through the ETP and partially used in gardening activities. Excess grey water was released through the public drainage system.

Flow of resources

This section covers the key findings from the case study. FPP directly applied the EMA methodology of MFCA to analyse the flow of materials, energies, water and waste. Although this was believed to be a straightforward application of the EMA framework, the exercise raised some new questions and generated learning that is contextual but could be generalized for the food manufacturing industry. The application of MFCA as part of the EMA method to FPP did not yield any surprises. As the manufacturing process at FPP was based on the principle of mass balance, MFCA was applied to evaluate waste output and ascertain the corresponding monetary loss. Sankey diagrams of materials, water and energy flow are enclosed (Figures 13.2–13.4).[1]

Since EMA supports organizational understanding of the waste chain and allows it to be optimized by bringing in the considerations of in-house recycling and other options, another unique insight could be achieved. Within the engineering industry, organizations can use the 5Rs (reusing, reducing, recycling, remanufacturing and reverse logistics) to lower waste output (Kakkuri-Knuuttila, Lukka and Kuorikoski 2008; Kasai 1999); however, FPP would not use waste pulp and peel in the production of fresh pulp as it would severely compromise the product's characteristics (pH, acidity, Brix value, colour and pigmentation) and result in contaminated products that would lead to unwanted losses. Accordingly, this points to the yield having a theoretical upper limit, which could be true for the entire food-processing sector. This

---

[1] For water, the unit of measurement has been converted to kilograms (kg) from litres (l) and for electricity to megajoules (MJ) from kilowatt hours (kWh) using standard conversion factors in these diagrams. The Sankey diagrams are drawn using STAN2 software (courtesy: Institute for Water Quality, Resource and Waste Management, Vienna, downloaded from http://iwr.tuwien.ac.at/resources/downloads/stan.html).

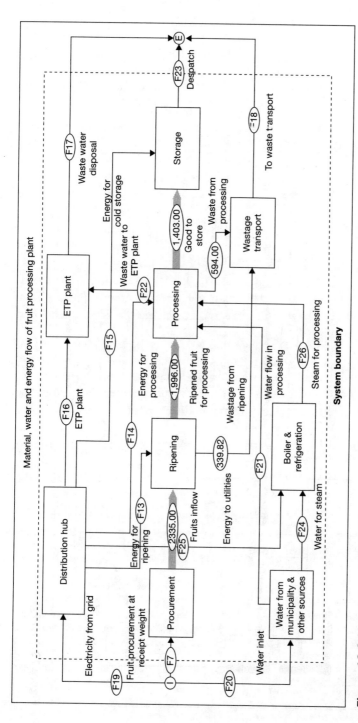

Figure 13.2 Materials and waste flow

Source: Author.

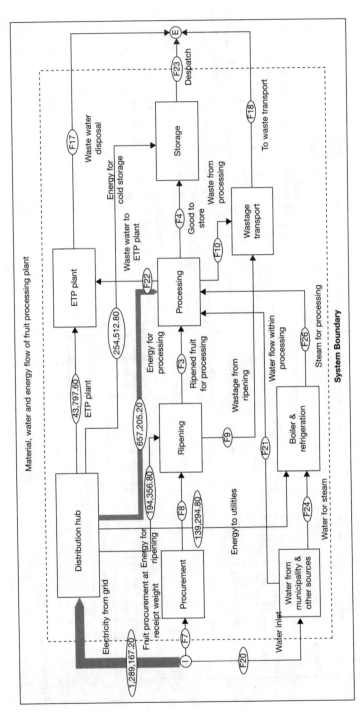

Figure 13.3 *Energy flow*

Source: Author.

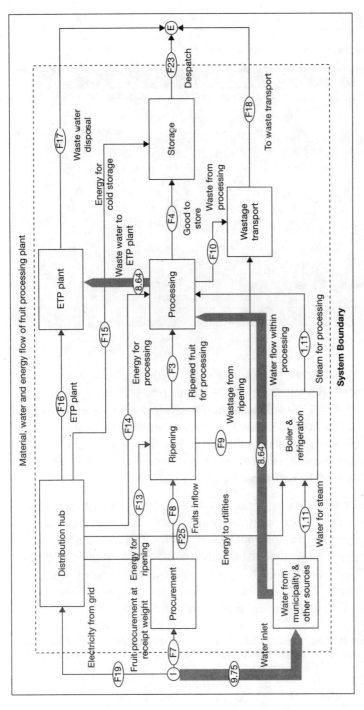

Figure 13.4 *Water flow*

Source: Author.

Table 13.2 *EMA computations for FPP*

As per EMA	As per the case study	Amount (in ₹)
Material wastes	Waste of 1,643 tonnes generated	53,772,237
Non-product outputs	Waste water of 21,117 kl	1,279,620
Waste and pollution prevention costs	– Cost of maintaining environmental certification – Running cost of environmental facility – Operational costs of other activities with environmental considerations	None
R&D expenditure	New initiatives for reducing environmental load	Not available
Less tangible costs	Emission externality of ~1.4 mtCO$_2$e per annum	316,575[2]
Total cost		55,368,432

Source: Author.

forces an acknowledgement that even with the deployment of the most advanced technologies, waste is an inherent part of any conversion process. Still, FPP could have sought ways to use the waste (for biogas, organic manure or compost) and reduce or nullify the corresponding environmental impacts, but it did not, and this could be related to the lack of environmental thinking. Accordingly, the EMA computations for FPP would be as shown in Table 13.2.

Since the maintenance of cost accounting records is a part of the statutory framework in India, the role of any alternative cost accounting exercise in practice would remain confined to that of an alternative computational method. Although FPP did not need to follow these rules as it did not meet the criteria that made it mandatory, the statutory nature of cost accounting anyway limits the role of EMA within the accounting fraternity, unless statutorily promoted.

[2] Emission costs at INR 225/tCO2e, or USD 4.5 at an assumed exchange rate of ₹50 = USD 1) (Ecosystem Marketplace and Bloomberg New Energy Finance 2011).

ENVIRONMENTAL ACCOUNTING FOR THE FIRM

Quantification of environmental aspects

To retrofit data from FPP, Table 13.3 has tabulated the operational data covering procurement (P1) and ripening and processing (P2–P3) operations. Data accumulated on a monthly basis helped in creating temporal accounts of the environmental aspects. The first step towards this was to quantify the aspects that were generated by the MFCA exercise. Emissions due to the transportation of raw materials covered procurement (P1) of raw or ripened fruits, including its transportation from orchards to the plant in trucks or lorries (8–11 tonnes per consignment), which would generate Scope 3 emission (as per GHG accounting), and were computed by using emission factors (GHGs 2,417.95 g km^{-1}; SO$_x$ 1.42 g km^{-1}, and PM 0.2 g km^{-1}) for 'Trucks and Lorries' in India (Ramachandra and Shwetmala 2009) for the distance travelled. Emissions due to energy use were based on the electricity sourced from the grid and consumed during processing (P2), which would generate Scope 2 emission, and were computed per the emission factor of 980 tCO$_2$e/ kWh (Central Electricity Authority 2011).

Valuation of environmental aspects

Next, the physical quantities from Table 13.3 were converted into the equivalent financial values. The solid waste is monetized using the social cost of disposal not internalized. The cost of solid waste disposal is ₹3,500 tonnes^{-1} as the uncovered cost was not internalized as part of the municipal services in Mumbai (as discussed previously). GHG emissions were valued at the market rate of USD 4.5 per tCO$_2$e (₹225 at an exchange rate of ₹50 per USD: Ecosystem Marketplace and Peters-Stanley et al., p. 27)—the average rate for Indian projects in the voluntary emissions credit market. Waste water was valued at the resource replacement rate of ₹50 kl^{-1}.

Table 13.3 Select environmental aspects of FPP

Months	Product type (Units)[6]	Transportation of fruits			Electricity		Waste	
		GHG (tCO2e)	SO$_x$ (tonnes)	PM (tonnes)	Consumption[3] (kWh)	GHG (tCO2e)	Waste water[4] (kl)	Solid waste[5] (tonnes)
May	M	238	0.14	0.02	160,436	157	4,269	246
Jun	B	4	0.002	0	197,666	194	4,731	680
Jul	B, P, G	37	0.022	0.003	116,399	114	1,857	142
Aug	P, G	23	0.013	0.002	108,286	106	2,550	122
Sept	B, G	86	0.05	0.007	151,423	148	3,115	280
Oct	M	0	0	0	83,931	82	1,450	55
Nov	G	20	0.011	0.002	75,377	74	1,554	72
Dec	X	43	0.025	0.004	82,625	81	1,591	90
Total		450	0.264	0.037	976,143	957	21,117	1,687

Scope 3 aspects

[3] Excluding ripening of banana (fresh).

[4] Production usage + tanker water.

[5] Excluding waste from banana (fresh).

[6] SO$_x$ = sulphur oxide emissions; PM = particulate matter, M—Mango puree and concentrate, B- Banana puree and concentrate, P- Papaya puree and concentrate, G- Guava puree and concentrate, X- Multi-fruit mix

Environmental ledger of aspects

The journalization and ledgerization process (Table 13.4a–b) used the double-entry system to transfer the aspects to the respective aspects accounts (of an asset nature) by debiting it, as these aspects are physical in nature and add to the natural asset pool of waste and emissions (same as finished products). The credit went to the respective environmental account (of a liability nature) to reflect the corresponding externalized liability. If any environmental aspect was not monetized (e.g., non-GHG emissions), it could still be accounted in the physical (or inventory) account only. The following accounting rules were followed in this process:

- The journal entries had to be balanced across all quantities and values.
- One accounting entry would use the same units of measurement (e.g., ₹, tonnes, kl, etc.)
- In case monetization could not be performed, only aspect accounting (in physical terms) would take place.
- To account for more than one aspect within a transaction, each aspect would have to be journalized.

Accordingly, the environmental ledgers were drawn (Table 13.4a–b).

Table 13.4a reflects the journalized entries of waste accounting for the period of observation, which resulted in a period-end balance of 1,687 tonnes of solid waste, creating externalities of ₹5.73 million on account of the disposal cost saved by relying on the municipal waste management system. However, Table 13.4b reflects the ledger account of product B, which resulted in an overall environmental load of ₹1.3 million. Table 13.4a–b present the combined t-accounts recording physical and monetary conversions together. Organizations can also maintain separate accounts to record aspect inventories and their monetary implications. As evident, the accounting entries maintain temporality and traceability for the underlying transactions. Also, non-GHG entries, like SO_x and PM, have been accounted physically, in the absence of a suitable monetary conversion. This reflects that the aspect quantities might not always be equated at the quantitative level (due to the non-convertibility of units). The externalities incurred or saved

Table 13.4a Solid waste (externality) t-account

Date	Particulars	Quantity (tonnes)	Debit (₹)	Date	Particulars	Quantity (tonnes)	Credit (₹)
05/xx	To env. liability a/c	246	861,000	12/xx	By Balance c/f	1,687	5,731,250
06/xx	To env. liability a/c	680	2,380,000				
07/xx	To env. liability a/c	142	497,000				
08/xx	To env. Liability a/c	122	427,000				
09/xx	To env. liability a/c	280	980,000				
10/xx	To env. liability a/c	55	19,250				
11/xx	To env. liability a/c	72	252,000				
12/xx	To env. liability a/c	90	315,000				
	Total	1,687	5,731,250		Total	1,687	5,731,250

Source: Author.

Table 13.4b Product B: Total environmental aspects t-account

Date	Particulars	Debit (₹)	Date	Particulars	Credit (₹)
07/xx	To env. liability a/c:		08/xx	By balance c/f	635,350
	Fruit travel (GHG: 239 tCO$_2$e)	53,775			
	Fruit travel (non-GHG: 0.02 tonnes)	–			
	Electricity (GHG: 95 tCO$_2$e)	21,375			
	Solid waste (125 tonnes)	437,500			
	Waste water (2,454 kl)	122,700			
	Total	635,350		Total	635,350
09/xx	To balance b/f	635,350	10/xx	By balance c/f	1,304,050
	To env. liability a/c:				
	Fruit travel (GHG: 53 tCO$_2$e)	11,925			
	Fruit travel (Non-GHG: 0.034 tonnes)	–			
	Electricity (GHG: 79 tCO$_2$e)	17,775			
	Solid waste (154 tonnes)	539,000			
	Waste water (2,000 kl)	100,000			
	Total	1,304,050		Total	1,304,050

Source: Author.

Table 13.5 Externalities incurred/saved due to the business activities of FPP

Type of activity	Environmental aspects	Savings	Value (in ₹)
Disposal of solid waste	Solid waste	Negative cost of 1,687 tonnes landfill (without any credit for waste utilization)	−5,731,250
Waste water released through drainage	Waste water	Negative cost of 11,000 kl waste water drained (assuming 50% used in gardening)	−550,000
Externality due to non-renewable energy	Emissions	Negative costs of ~1.4 tCO2e from energy and transport	−316,575
Total			−6,597,825

Source: Author.

as per EMA as well as by using environmental accounting ledgers are tabulated next (Table 13.5).

From the findings, it could be concluded that the selected case helped to explore the feasibility and suitability of the EMA methodology and in generating management information on the environmental aspects. The fruit yields computed by MFCA were in line with FPP's MIS, and the financial loss on waste could generate additional information for the management, albeit of little use. Moreover, the statutory nature of cost accounting in India is an impediment for EMA to evolve as part of the management accounting practice. However, the insights gained from the study help us to highlight that the environmental impact of business activities is not a direct result of the generation of aspects, but of how the aspects are discharged or disposed of in the environment. If FPP considered leveraging an external recycling mechanism for its solid waste or reused treated waste water, its environmental impacts would have significantly reduced. The second case study in the next chapter explores these findings against a different contextual background.

Environmental Accounting: Part II

INTRODUCTION

The experiment from the previous chapter is being continued here to explore further and reflect on how environmental accounting of aspects can be an effective way to manage these aspects and to improve the accountability of firms, especially if a firm is caring for the environmental impacts it is producing. This chapter explores these ideas through another case study before synthesizing the learning of these two chapters to conclude why and how environment accounting could help firms deal with aspects and impacts in quantitative terms, even with the additional overhead of accounting for it. Accounting automation can, however, effectively handle the book-keeping process and more of it has been discussed in the next chapter. Learning from these case studies (in this chapter and the previous one) brings practical perspectives to bear on the concepts discussed in the literature and the diversity that exists within the theories. While FPP, owing to its manufacturing background, could adapt to the computational arrangement of environmental aspects at the process level, CHS—the subject of our second case study—could hardly connect the multiple chains of activities to the intangible nature of hospitality services, as we shall see. Learning from these case studies is backed up by a discussion of how the accounting of externalities would need to assimilate fragments from different theories.

CASE DETAILS

The rationale for the second case study was to generate a view of environmental accounting in an industry where mass balance is not the primary method to establish an input–output link—in a service organization. A service organization takes a different view of the stock and flow of environmental aspects, which has not been covered in the literature much. Interested readers can refer to Debnath (2015) for more details. The following section summarizes the study and its findings.

Project site

The second case proved the relevance of EMA to the hospitality sector in general and expanded the knowledge base with the findings. The study covered two co-located hotels (a five-star and a three-star) in India, managed by a hospitality services firm (CHS). The study followed the same methodology as case one. First, the site was visited and the need for the study was explained to the management. This helped the firm to nominate departmental contacts. This was followed by the gathering of information on the workings of CHS and its processes through interviews with the departmental contacts, which were conducted over a period of three months and took 10–12 visits at regular intervals. Questions related to the workings of the facilities and processes were covered through these unstructured interviews. Thereafter, operational data was gathered at quarterly intervals to cover the operational performance for a year. As in case of the first case, the generalizability of the research was maintained by confining the study within the area of interest. Accordingly, the study focused on using EMA artefacts to explore the data collected from the site to reflect on the environmental considerations of the firm.

The service environment

Both the CHS facilities were full-service business hotels and offered lodgings, boutique restaurants, bars and lounge facilities to business travellers, along with banquet arrangements and conference halls. The guest-service life cycle covered room reservations and check-in, followed by stay, boarding and checkout. The guest amenities and services consumed materials, water, energy and other resources, and produced waste. Like in the first case, a mix of arbitrary and market rates were used as financial proxies to translate the physical data into financial equivalents wherever required. The site visits helped to map the processes and connect the collected data to the stock and flow of materials and services, and thence to compute the environmental aspects. The outline of the service environment (gate-to-gate cycle) is detailed in the block diagram in Figure 14.1. Being Ecotel-certified—a certification indicating environmental and social leadership in the hotel and hospitality business (UNEP 2012)—both the facilities were equipped with infrastructural and operational arrangements to provide environmentally conscious hospitality arrangements, as discovered from the interviews and verified by the environmental handbook of the facilities.

Analysis of key findings

Before detailing the quantitative performance of CHS, the findings from FPP are validated in this segment. As far as the application of MFCA (as a part of the EMA method) to CHS was concerned, as well as the use of mass balance to evaluate costs and flow of waste, CHS being a service-based organization, MFCA could not be applied as the services did not follow the mass balance principal. Also, the cost of waste was not derived as a yield and operational losses were of little concern to CHS. Accordingly, the waste reporting method that was used here was based on cumulative waste quantities. In any case, the internal cost of waste could not help the management as it could not relate it to the cost of services due to the lack of a causal relationship; neither was recycling of food and packaging waste feasible, as explained below. Similar to the FPP, the generation of solid waste and waste water were tracked by CHS as part of its existing MIS reports. At the contextual level, the statutory cost accounting rules applicable to Indian organizations were not mandated for CHS, as the service industries are outside the purview of the legal statute of cost accounting. So CHS had no incentive to maintain the cost sheets other than for internal decision-making purposes. As far as waste handling was concerned, CHS transferred its biowaste to an in-house vermicomposting facility and waste water to the community ETP for recycling. Vermicomposting generated a bio-fertilizer compost that was sold

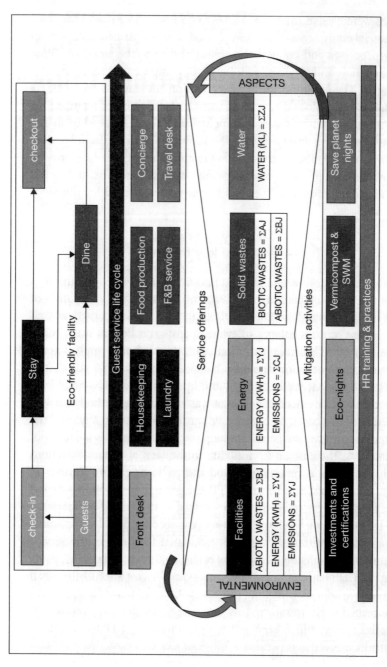

Figure 14.1 Operational layout of CHS

Source: Author.

at a nominal rate, whereas treated grey water was received back and recycled. CHS had a specific layout of pipes to circulate grey water and used it for the designated purposes. Other types of waste were reused, while the ones that could not be reused or recycled (butter paper, oil cans, etc.) were sent to landfill (Tables 14.1–14.2).

Considering the complete recycling of biowaste and reuse of grey water, CHS saved the environmental impacts that would have been generated if the aspects were disposed of in the conventional manner. This in turn resulted in the saving of the social costs of waste disposal. As part of its social commitments, CHS also participated in programmes that would reduce community waste and save social costs. These tangible savings could not be incorporated as part of the present EMA construct, however. As a contrast to these practices, CHS had outsourced its laundry services, where the waste water could not be reflected as part of the EMA construct, as it is yet to incorporate supply-chain impacts as part of the construct. Accordingly, environmental accounting for the CHS study is as follows (Table 14.3).

ENVIRONMENTAL ACCOUNTING OF ASPECTS

To analyse the robustness of the EA exercise from the first case (FPP), it was is repeated for CHS for the equivalent period and by tabulating the environmental aspects in the same way (Table 14.4). The valuations of the aspects are held at the same level as the first case, except for solid waste. As CHS did not transfer solid waste to the public domain but converted it into compost using an in-house vermicomposting facility, it generated a positive externality that is reflected as credit entries representing a savings in social costs of ₹3,500 tonnes^{-1} (Table 14.5a–d).

Accordingly, the environmental ledgers were drawn up. The credit balance of Table 14.5a reflects the social externality saved due to vermicomposting of biowaste in the form of an equivalent cost of MSW management. Table 14.5b brings the supply-chain effects into the books of the beneficiary, in the form of resource accounting for the waste water of outsourced laundry, accounted as an environmental liability of CHS. Table 14.5c reflects the equivalent carbon value of emissions. Accordingly, the environmental liability at the end of the

Table 14.1 Periodic consumption of resources in respective units[1]

Month	Guest nights	Electricity (kWh)	Water (kl)	HSD (litres)	F&B (covers)	Linen (par)	Wet garbage (kg)	Dry
Apr	3,550	347,783	4,156	141	7,997	43,555	11,705	3,068
May	3,128	353,394	5,140	141	7,537	54,585	11,483	2,635
Jun	3,038	332,496	5,899	82	5,283	38,164	8,050	2,287
Jul	3,290	352,577	6,356	92	5,585	42,058	8,349	2,508
Aug	3,299	368,708	4,906	27	7,042	57,657	8,212	2,926
Sep	3,643	357,025	5,632	76	8,988	48,628	8,098	2,876
Oct	3,848	380,822	4,344	228	7,330	61,056	8,423	3,051
Nov	3,594	341,721	4,041	76	6,930	50,238	8,114	2,603
Dec	3,421	347,388	8,956	139	13,434	55,864	8,090	1,156
Jan	4,200	516,479	10,362	16	14,368	57,239	8,086	2,455
Feb	3,998	459,976	9,495	79	13,467	61,428	7,561	2,833
Mar	3,959	520,641	8,862	549	11,974	53,466	8,291	2,522
Total	42,968	4,679,010	78,149	1,647	109,935	623,938	104,462	30,920
Per guest night		108.90	1.82	0.04	2.56	14.52	2.43	0.72

[1] par = set of linen on hand; F&B = food and beverage; HSD = High speed diesel.

Table 14.2 Environmental aspects from resource consumption (with units)

Month	Guest nights	GHG due to electricity (tCO$_2$e)	Water consumption (kl)	Water consumption due to linens wash (kl)	GHG due to HSD (tCO$_2$e)
Apr	3,550	341	11,325	7,170	0.42
May	3,128	346	13,647	8,507	0.42
Jun	3,038	326	11,674	5,775	0.24
Jul	3,290	346	12,772	6,416	0.27
Aug	3,299	361	13,821	8,916	0.08
Sep	3,643	350	13,075	7,443	0.23
Oct	3,848	373	13,772	9,428	0.68
Nov	3,594	335	11,803	7,761	0.23
Dec	3,421	340	17,553	8,597	0.00
Jan	4,200	506	18,984	8,621	0.0
Feb	3,998	451	18,875	9,381	0.0
Mar	3,959	510	17,024	8,163	0.0
Total	42,968	4,585	174,326	96,177	2.6
Per guest night		0.107	4.057	2.238	0.073

Table 14.3 *EMA computations for CHS*

As per EMA	As per the case study	Amount (in ₹)
Material wastes[a]	Dry and wet garbage: bottles, packing materials, empty containers, food wastes and others (100% recycled) 150 tonnes per annum	0.00
Non-product outputs[b]	Waste water (100% recycled) ~45,000 kl per annum	0.00
Waste and pollution prevention costs[c]	• Fixed costs per annum of maintaining Ecotel certification	250,000[a]
	• Running cost of vermicomposting facility	+ 60,000[b]
	• Operating cost of other activities with environmental considerations	Unascertainable
R&D expenditure[d]	New initiatives for reducing environmental load	Not available
Less tangible costs[e]	Emission externality of ~4.5 tCO2e per annum	1,015,000[c]
Total cost		1,325,000

Source: Author.

Notes: a. Cost of Ecotel Certification.
b. Assumed maintenance cost of vermicomposting facility (1 person at ₹5,000 per month).
c. Emission costs at ₹225/tCO₂e (USD 4.5 at assumed exchange rate of ₹50 per USD) (Ecosystem Marketplace and Bloomberg New Energy Finance 2011).

Table 14.4 Analysis of waste generated

Month (2012)	Food production (covers)	Garbage (tonnes)	Linen for wash (par)	Water consumption (kl) Supply	Water consumption (kl) Laundry outsourced	Water consumption (kl) Total	Energy (kWh)	GHG (tCO₂e)
Apr	9,069	14.8	43,555	4,156	7,170	11,326	347,783	341
May	8,248	14.1	51,949	5,140	8,507	13,647	353,394	346
Jun	5,283	10.3	35,045	5,899	5,775	11,674	332,496	326
Jul	5,585	10.9	39,003	6,356	6,416	12,772	352,577	346
Aug	7,586	11.1	54,359	4,906	8,916	13,822	368,708	361
Sep	10,111	11.0	45,280	5,632	7,443	13,075	357,025	350
Oct	8,144	11.4	57,461	4,344	9,428	13,772	380,822	373
Nov	7,790	10.7	47,311	4,041	7,761	11,802	341,721	335
Dec	13,434	9.2	52,360	8,956	8,597	17,553	347,388	340
Total	75,250	103.7	426,323	49,431	70,012	119,443	3,181,914	3,118

Source: Author.

Table 14.5a Solid waste (externality) t-account

Date	Particulars	Quantity (tonnes)	Debit (₹)	Date	Particulars	Quantity (tonnes)	Credit (₹)
12/xx	To balance c/f	103.7	362,950	04/xx	By env. liability a/c	14.8	51,800
				05/xx	By env. liability a/c	14.1	49,350
				06/xx	By env. liability a/c	10.3	36,050
				07/xx	By env. liability a/c	10.9	38,150
				08/xx	By env. liability a/c	11.1	38,850
				09/xx	By env. liability a/c	11.0	38,500
				10/xx	By env. liability a/c	11.4	39,900
				11/xx	By env. liability a/c	10.7	37,450
				12/xx	By env. liability a/c	9.2	32,200
	Total	103.7	362,950		Total	103.7	362,950

Source: Author.

Table 14.5b Waste water t-account

Date	Particulars	Quantity (kl)	Debit (₹)	Date	Particulars	Quantity (kl)	Credit (₹)
04/xx	To env. liability a/c	7,170	358,500	12/xx	By balance c/f	70,012	3,506,000
05/xx	To env. liability a/c	8,507	425,350				
06/xx	To env. liability a/c	5,775	288,750				
07/xx	To env. liability a/c	6,416	320,800				
08/xx	To env. liability a/c	8,916	445,800				
09/xx	To env. liability a/c	7,443	372,150				
10/xx	To env. liability a/c	9,428	471,400				
11/xx	To env. liability a/c	7,761	388,050				
12/xx	To env. liability a/c	8,597	429,500				
	Total	70,012	3,506,000		Total	70,012	3,506,000

Source: Author.

Table 14.5c *Emissions t-account*

Date	Particulars	Quantity (tCO₂e)	Debit (₹)	Date	Particulars	Quantity (tCO₂e)	Credit (₹)
04/xx	To env. liability a/c	341	76,725	12/xx	By balance c/f	3,118	701,550
05/xx	To env. liability a/c	346	77,850				
06/xx	To env. liability a/c	326	73,350				
07/xx	To env. Liability a/c	346	77,850				
08/xx	To env. liability a/c	361	81,225				
09/xx	To env. liability a/c	350	78,750				
10/xx	To env. liability a/c	373	83,925				
11/xx	To env. liability a/c	335	75,375				
12/xx	To env. liability a/c	340	76,500				
	Total	3,118	701,550		Total	3,118	701,550

Source: Author.

Table 14.5d Environmental liability t-account

Date	Particulars	Quantity (tonnes)	Debit (₹)	Date	Particulars	Qty. (tCO₂e)	Quantity (kl)	Credit (₹)
12/xx	To solid waste a/c	103.7	362,950	12/xx	By waste water a/c		70,012	3,506,000
12/xx	To balance c/f		3,844,600	12/xx	By emission a/c	3,118		701,550
	Total	103.7	4,207,550		Total	3,118	70,012	4,207,550

Source: Author.

period represents the monetized balance (in quantity and monetary terms), reflecting the externalities not annulled by the organization (Table 14.5d). As the scope of this research is to reflect the role of accounting artefacts in the information generation process, secondary accounts of environmental aspects are assumed to be the logical follow-through of the process. Accordingly, this has not been delved into any further and could be taken up as part of future research.

As compared to FPP, the CHS facilities could be calibrated against the externalities that it incurred or saved as part of its operations. The performance of the CHS facilities would also indicate the social costs it saved for society, which were captured within the environmental accounting constructs. This could also enable inter-firm comparisons of environmental performances within the same industry and included the saving on SWM services to clean up the community waste (Table 14.6). The waste water generated due to the outsourced laundry is a waste of a natural resource, as commercial laundries drain the waste water instead of recycling. This is a negative externality and was considered as part of the environmental performance of CHS facilities. By

Table 14.6 *Externalities incurred/saved due to the business activities of CHS*

Type of activity	Environmental aspects	Savings	Value (in ₹)
Vermicomposting	Solid waste	Saved 104 tonnes of wastes from landfill	+362,950
Composting of floral wastes from festivals	Solid waste	Saved 10 tonnes (assumed) floral wastes composted	+35,000
Outsourced linen washing	Waste water	Negative cost at replacement rate of water for ~70,000 kl per annum	−3,506,000
Use of non-renewable energy sources	Emissions from energy use	Negative costs of ~3.2 tCO_2e	−701,550
Total			−3,809,600

Source: Author.

excluding an in-house laundry facility, CHS could overstate its environmental performance as compared to others but this also resulted in a cost of ₹8.75 million per annum (at the resource replacement rate). Based on the selective performance data and its accounting, these externalities created an environmental obligation of around ₹3.8 million for 26,000 guest nights or ₹147 per guest night, which could now become a part of the corporate considerations in the environmental accounting construct. However, the hotel management also interacted with schoolchildren from around the locality and organized workshops with other business facilities to impart to them practical tips on being environmentally friendly. These activities of the firm generated positive externalities, but would need suitable valuation methods to find a place in the proposed scheme.

Findings from CHS (as compared to the FPP) reflect the shift in industry context (from manufacturing to hospitality services) and the dilution in the quantitative evaluation of costs that would otherwise offer limited benefits in analysing the cost of waste. On the other hand, the new findings of positive externalities and contribution to social costs were unique to the firm but could be generalized as outcomes of pro-environmental strategies of businesses. This included the key service areas that could be outsourced by firms where the impacts would remain outside the traditional construct. In that sense, the findings from CHS expanded the learnings from FPP and added richness to it. At the same time, by following the same methodology to conduct the study, the generalizability and repeatability of the adopted methodology could be tested. The findings further establish that the management of an organization can improve care for the environment if it can be related to competitive positioning within the industry and compared to how other firms are handling the contemporary environmental situation.

CASE STUDY III: VALIDATING ENVIRONMENTAL ACCOUNTING IN AN UNRELATED ENVIRONMENT

To test the generalizability of environmental accounting beyond proximate research, a third case study was conducted to validate the framework in an altogether different environment and demonstrate its capability of handling a diversity of industry, aspects and time. This case was an ex-post study and used historical data from one of the information system projects.

Background

A system implementer DEF Ltd bagged a 'procure-to-pay automation project' contract (Project A hereafter) and rolled it out to the India division of ABC as a pilot. The pilot project involved a team of 12 consultants (two US-based and the rest in India). From the client side, the project was spearheaded by two representatives from the UK office (as part of project management office, or PMO) and four subject matter experts from the US. This case study is limited to covering the environmental aspects generated during this project due to the project members travelling from different locations to India at different points in time. To keep the study focused, the project team's emission inventory generated by international air travel is the only environment aspect explored in the study.

Computation of environmental aspects

Using data from the organizational records, environmental records were created for the project members' air travel, by converting the information into an equivalent tonnage of CO_2 (tCO_2e) and accounting it against the pro-rata billing cycle, which was based on the billing estimate of USD 421,000. The travel details for the project team during the relevant period are as given in Table 14.7.

Valuation and accounting of environmental aspects

To compute and account for the aspects, the gold standard of an average €15 per tonne of carbon equivalent has been taken as the offset to monetize the emissions (IATA 2008). The result reflects that the project generated an incremental environmental load of 77 tCO_2e over 412,000 kilometres of travel, equivalent to a €1,000 in terms of abatement value, which remained outside commercial and accounting

Table 14.7 Travel miles and emissions produced during Project A

Milestone	Start date	End date	Consultants	Subject matter expert	PMO representatives	Travel distance (km)	Carbon emission (tCO$_2$e)
Conference room pilot I (CRP I)	12 Aug	16 Aug	1	1	2	78,840	15.89
Conference room pilot II (CRP II)	12 Sep	19 Sep	0	0	0	0	0
System integration testing (SIT)	06 Nov	19 Nov	2	3	2	166,340	30.33
User acceptance testing (UAT)	04 Dec	17 Dec	1	3	2	166,340	30.33
Total						411,520	76.55

Source: Author.

considerations (Table 14.8). Although the scale of the project was small and the people moving between different countries were a few, it could be asserted that technology projects are not in general environmentally benign, not to mention the environmental aspects due to electricity use, stay in hotels, car travel, and so on, that are not considered due to lack or inaccessibility of data. If the project had been for an organisation in an industry from a non-Annex B region per the Kyoto Protocol, it would have brought in additional considerations of arranging for carbon commitments as part of the project, being researched. As such, the current project management practices are yet to evolve beyond the regular costs and accrued revenues (Curkovic and Sroufe 2007). Accordingly, it needs environmental accounting to capture the environmental aspects of the project activities.

SYNTHESIZED LEARNING FROM THE CASE STUDIES

These case studies evidence that the accounting language is capable of handling multiple units of measurement as part of a system and of generating accounting ledgers in physical and monetary terms. The uniformity of procedures seen in this segment establishes the feasibility of an environmental accounting framework to handle environmental aspects in general and establishes the generalizability of the overall process. It also supports the repeatability across cases. Further, the learning synthesized from the cases is detailed and leads to new knowledge.

Solid waste and its disposal

Waste is an inevitable by-product of any production process that is usually discarded but could be reduced to an extent by using technological improvements. The internal costs to generate and quantify waste can offer more information to the management and could lead to improvements in operational arrangements, but information on environmental externalities and social costs would typically remain outside the firms' accounting practice. From the contextual perspective of these cases, it appears that in the absence of any specific regulations on waste disposal by commercial establishments

Table 14.8 T-account of emission aspects due to travel miles in Project A

Date	Particulars	Quantity (tCO₂e)	Debit (€)	Date	Particulars	Quantity (tCO₂e)	Credit (€)
08/xx	To env. liability a/c	15.89	238.35	12/xx	By balance c/f	76.55	1,148.25
11/xx	To env. liability a/c	30.33	454.95				
12/xx	To env. liability a/c	30.33	454.95				
	Total	76.55	1,148.25		Total	76.55	1,148.25

Source: Author.

or industries, other than those categorized as 'hazardous' under the Hazardous Wastes (Management and Handling) Amendment Rules of 2000 in India or as 'hazardous' or 'special' categories of waste by the EPA (2010), or other similar legislations in the respective countries, organizations might depend on the community disposal mechanisms to dispose of waste, so long as the cost of using these facilities is commercially viable. In other words, as long as using municipal disposal facilities is cheaper than developing alternate mechanism of waste disposal, businesses will not shy away from using the municipal facilities. However, the municipal waste management system incurs social costs that might not always get equitably distributed, not to mention the externalities that are expected to be borne by the environment and society, which remains a methodological concern for EMA. Although developed countries have internalized the costs associated with public waste disposal using different market mechanisms (Jasch and Lavicka 2006), these solutions are yet to result in an overall decline of waste levels or motivate businesses to be efficient and effective. Moreover, the institutionalized practice leads to a 'right to waste' that can be purchased for a fee and can prove to be counterproductive in the long run, yet to create a permanent change in the attitudes of firms.

Conscientious corporate citizens such as Interface Inc. (carpet/tile manufacturer), Patagonia and 3M have developed sustainable waste-reduction techniques as part of a cradle-to-cradle life cycle and other initiatives (Esty and Winston, 2006). Others may follow suit if the organizations are made aware of or held accountable for the externalities they generate. Environmental accounting is expected to help firms improve their overall approach towards environmental care by providing a mechanism that can directly reflect their performance in real time. Accordingly, environmental accounting can help firms to develop insights into their decisions (e.g., externalized liability for CHS due to landfilled waste that needs a permanent cure) by generating temporal information that remains tied to the source activity or business transaction.

Waste water

Processing of waste water through local or site-specific ETPs is a common practice for firms, before releasing the water to the public drainage system, which would release the treated water to the water bodies. However, this has not always been the case, for example, in the case of developing and underdeveloped countries, where waste water could be released directly to the water bodies, resulting in a severe loss of water resources. At the same time, waste water has not always been reused by developing in-house or community-level grey-water recycling systems, where treated water could find uses other than for drinking or human consumption. Cities in developing countries are yet to develop the public infrastructure needed for this endeavour. This lack of infrastructure is in turn due to lack of legislation to promote grey-water recycling as a mandate for the community. With depleting levels of potable water globally and an ever-rising population, the market mechanism to price such a scarce resource would hardly improve the situation in the long run. Connected to this is the apathy of the industries who, rather than developing a community ETP to treat waste water, simply release it to public water bodies, a common behaviour in the industrial parks in developing countries that is adding to the water crisis. Add to this the cost of recycling water that is not factored in by municipalities everywhere (Mahadevia and Wolfe, 2008). The intent of *accounting* for waste water is to help firms appreciate the stress that they are contributing to within a region, which could lead to business disruptions as well. This new-found insight could help firms to participate in improving the disparity within the region and be recognized as valued members of the community they serve, as hospitality services of some global chains have started resorting to.

Emissions accounting

The emission of GHGs and several non-GHGs has harmful effects on the biosphere and contributes to changes in climatic conditions. While GHG accounting offers a way to inventorize emissions, there is no integration of emissions stock within the EMA framework. In the absence of suitable incorporation of emissions within the environmental accounting process, the decisions to move from energy-intensive machinery, equipment and processes to low-energy ones (e.g., the installation of a bagasse-based boiler in FPP instead of an oil-fired boiler) would depend mainly on the economics of decisions (lowering cost per kg of steam), with no or low regard for environmental benefits. On the other hand, the use of technological solutions to reduce the use of energy (e.g., the use of glycol chiller technology, automatic water-level controllers and timer-based energy devices, amongst others) could result in optimized use of energy. Still, carbon neutrality is something that firms need to consider, and environmental accounting can help firms contribute to emission accounting, instead of forcing firms to act through legislative changes. The argument here is that the environmental impact of a business decision can be evaluated better if it can be tied to performance levels (before and after the effective change), so that temporality becomes a natural requirement for environmental accounting to handle and extend its capability to cover all types of waste and externalities.

Approaches to improving environmental care

As compared to EMA, where insights into environmental performance are limited within the organizational boundary of the firms, environmental accounting offers information on the types of environmental aspects and how these are being handled. For example, the solid waste account of CHS details saved social cost as an asset, whereas waste water from outsourced laundry is registered as a liability that CHS should be considering, with improved commitment to remaining a pro-environmental business as it has declared itself to be. Same is the case of emissions, which CHS should be taking care of. These insights would not have been possible without letting environmental accounting break down the boundary of traditional accounting and reflect how firms are contributing to the environmental duress. Although this might not resolve differences that are inherent to the industries in regard to how they operate and/or use resources (e.g., discrete manufacturing versus the hospitality business versus the mining sector), information generated and disseminated through environmental accounting could still institutionalize a shared vocabulary, which is the need of the hour, including devising common terminologies to express how firms might be viewing their performance as compared to others or for analysing industry-specific trends (e.g., environmental care institutionalized by the extractive industry through its operating norms). The framework also expands beyond the boundary of known aspects and handle areas that lack computational insights, such as emissions of non-GHG gases (like F-gases) or the positive social externalities that CHS generated through the workshops it conducted for other firms to become environmentally friendly, and this could be pursued as a part of future research.

The two cases discussed also present a contrasting approach towards environmental care. At a theoretical level, between these two cases, we covered firms that are at the initial stages of developing environmental performance as one of their goals (case 1) to the other extreme where the firms have managed to develop environmental considerations as their competitive edge (case 2). Whereas in case 1, we witnessed a firm yet to plan for bringing environmental considerations into its operations, the firm in case 2 firm has developed an entire infrastructure and operational arrangements to incorporate environmental considerations within the business. In that sense, FPP could be classified as an organization with no or minimal environmental considerations, while CHS has risen to the morphogenetic level, where its environmental outlook has changed its business conduct (Fraser 2012). It is worth noting that Ecotel certification provided CHS with the necessary legitimacy and impetus to help its environmental care emerge as a competitive differentiator.

From the study, it was found that CHS followed a different mechanism for waste disposal (including for waste water), which limited the environmental impacts. However, unlike the engineering industry, where the organizations can use the 5 R's (reuse, reduce, recycle, remanufacturing, and reverse logistics), CHS could not recycle its waste to improve resource efficiencies in F&B services. This leads to the argument that waste is inherent to any process and in-house recycling has its own limitations. Such insights reveal that the environmental impacts of waste would depend on how the aspects are discharged into the environment. This is where accounting for externalities becomes an improvement to support sustainability. Although the generation of environmental aspects is inevitable in any physical process, to deal with them effectively and efficiently is the goal and also the measure of a firm's performance with regard to it. The possibilities from a firms' perspective are to ignore it altogether (business as usual), treat these aspects to improve the bottom lines (picking the low-hanging fruits), or take a wholesome approach to minimizing the firm's impacts on the environment (ethical approach). The role of accounting in this regard is to help businesses change the representations from stakeholders' needs to how these concerns are being taken care of.

Environmentally sensitive decision-making

FPP collected information on waste and developed metrics on internal efficiencies with regard to fruits and consumption of resources, whereas CHS used a similar set of data to monitor the environmental and business performance. FPP has SAP R/3 ERP software to capture data on its operations, whereas the data collection at CHS was rudimentary at best and maintained in Excel worksheets. Even with this seemingly odd inversion of technological advancement, the decisions are prioritized at CHS to be environmentally benign. The comparative positioning changes the paradigm of decision-making in these organizations. In the classical sense, FPP would explore decisions as formulated by MCDM:

$$\text{Min } Z = \sum C_{ij} \text{ (Minimize costs over i activities and j resources)} \quad (14.1)$$

However, in CHS, the decision-making framework adds another additional objective criterion:

$$\text{Min } Z_1 = \sum E_{ij} \text{ (Minimize environmental impacts over the same set)} \quad (14.2)$$

Methods like goal programming, analytical hierarchical programming (AHP) or MCDM techniques could be used in such situations to trade-off on different alternatives. The environmental outlook of CHS was based on the operational decisions built on minimizing (or nullifying) negative environmental impacts ($\sum E_{ij} \rightarrow 0$). However, in business terms, it shifted the decision-making paradigm completely. For example, the CHS facilities did not use wooden hangers (to reduce load on natural resources) and instead used ones made from other natural materials (e.g., cardboard hangers made from recycled paper or made of medium-density fibreboard, or MDF). Obviously, the price and life of the eco-friendly hangers are not comparable to that of the wooden ones, and neither are the environmental benefits contributed by the former are comparable to the cost differentials. This renders the classical decision-making paradigm of comparing the opportunity cost of environmental care against the economic benefits redundant and emphasizes the need for subjective criteria to go beyond the isolated identification of costs and to consider the environmental impacts of the business activities in a holistic manner.

In summary

The case studies offered a first-hand overview of how business activities generate environmental aspects and result in social and environmental externalities. Experimenting with multi-case strategy resulted in:

- Analyse the crucial links that formed part of the environmental considerations for firms.
- Study the repetition of underlying phenomena over multiple reporting periods, since ensuring the repetition of observations is a major issue in social experiments.
- Establish a chain of evidence by recording observations and verifying sample data from organizational artefacts.
- The variations in using different EMA methodologies to evaluate the outcomes in relevant chapters in and these case studies helped in placing the findings against the theory.
- Reflect on the limitations of using EMA methods, with their specificity to the manufacturing industry.
- Identify the limitations of EMA in these cases, where it could not reflect the savings in costs that CHS generated by recycling solid waste or the lowered social costs from better use of resources. This shows that the EMA computations are only a step away from the economic viewpoint.
- Emphasize the environmental embeddedness of firms. For example, we can view FPP as being representative of a class of

organizations where environmental leadership is yet to emerge and which are yet to develop sensibility towards the environment and its resources. In contrast, CHS can be viewed as a firm that has raised its environmental superiority beyond the certification level and established it as a competitive differentiator. Its care for societal and in-house biowaste could be a learning for others. At the same time, the outsourcing of resource-intensive laundry services sounds a cautionary note on the importance of having a system that can track resource utilization beyond the organizational boundaries.

It is natural to be critical of the risks associated with the exploratory nature of the environmental accounting framework and its underdeveloped rules, but its standardization can be taken up at the national or supranational levels and a standardized interpretation can be developed across firms and/or industries. In the ever-complex world of business, the overall performance of an organization would seldom be judged by its financial performance alone in the near future. The proposed framework is expected to bring the other dimensions of organizational performance within the equation and help firms follow sustainable thinking as a natural corollary of accounting.

Accounting Sciences and Sustainability Theories: Managerial Implications and Recent Advances

Environmental Accounting and Managerial Implications I

Carbon Accounting

INTRODUCTION

This chapter continues where the previous one left off to explain how any form of environmental accounting would need to support the information needs of the management and help firms in their quest to improve environmental performance. Following from the accounting viewpoint that has been presented in previous section, this chapter proposes some constructs that are independent, nascent and experimental in nature, but are based on sound business and accounting principles. These ideas include the development of an integrated view for waste and emissions (a quantitative construct that brings these two waste streams together) and the feasibility of developing a carbon accounting construct as an extension of the corporate environmental accounting framework. Other than experimenting with these diverse ideas, this chapter discusses the relevance of environmental accounting as a common denominator, establishing that the management could gain from the improved information flow and relating that to environmental performance.

EMISSION TRADING AND CARBON ACCOUNTING

Carbon accounting, which has been briefly discussed in Chapter 9, relates to a specific area of environmental accounting and deals with the measurement, accounting and reporting of GHGs released or sequestered through organizational activities. Moreover, carbon accounting within a region covers changes to its GHG profile due to economic activities, changes in the assets that are carbon sinks or sources of sequestered carbon (such as changes to forest cover or operations of the timber and lumber industries), and land use and land use changes and forestry (LULUCF)—this is a separate area of study and out of scope here. At the macro level, regional fluctuations in emission from material flow cycles can be aggregated and studied by implementing a green national accounting system, which has evolved into a separate discipline from being an area of study under environmental economics—field of study to understand the impacts of economic activities on the proximate environment. Although theoretically the aggregation of micro-level activities should be related to the macro-level changes, there is a dearth of studies to connect the two beyond economics, primarily due to multiple areas of study remaining disconnected and advancing independently. As a result, carbon accounting is currently being researched as a part of environmental accounting for firms, to offer a methodological advancement in recording the generation, trading and accounting of emissions. Moreover, this could improve decision-making tools that managers and experts can leverage to bring scientific temperament into the decision-making and reporting process. At a macroeconomic level, information from firms needs to be aggregated to develop emissions profiles of the different industries, which should lead to a better quality of forecasts that political and regulatory systems need to work towards conforming to local and global expectations.

Having said that, coverage of carbon accounting here is more from microeconomic perspectives, discussing how firms could improve their response on emissions and considering the new research in this field that can help firms respond better. With only the one accounting guideline that ever tried to respond to this ((International Financial Reporting Interpretations Committee, now referred to as IFRC Interpretations Committee, IFIRC3) having been scrapped later, firms are left to try different options and handle emissions and emission rights-related assets on their own. This is where we need to break from the traditional accounting theories to expand and connect to the changing world of business. As businesses in some regions have been granted initial emission rights, and firms can purchase and sell emission rights in the market, there is a gap that firms are finding difficult to deal with, because emissions-related accounting implications are not being considered holistically. However, the literature is yet to bridge the different areas of scholarship and offer any prescriptive solution to cover carbon and GHG issues in their entirety. Emissions are considered an inherent part of the technosphere that is the driving machinery of anthropocentric activities. Accordingly, for firms to own responsibility for them, especially in the absence of a regulatory framework, would need encouragement. Even if the firms would opt for it, they would do so to improve the associated economics—for example, the rising numbers of green buildings, which seems to suggest that these are carbon-neutral as well. There are some companies as well who claim that they are carbon-neutral (for example, ITC in India), although there is insufficient standardization to validate such a claim.

GHG accounting was the first attempt to help firms with their need to measure, account for and report GHGs generated, accounted and sequestered for the firms to move towards being carbon-neutral. However, GHG accounting was related more to managing the physical stock of emissions and depended on LCA to develop a stock and flow model. It divided emissions into different categories, depending on how these are produced due to the different activities of a firm. It recommended methodologies proposed by the Intergovernmental Panel on Climate Change (IPCC) for measurement of emissions that are applicable to a firm and adoption of the ISO 14064 standards. Outputs from this exercise can be used for accounting and reporting as well, for example, in GRI-compliant, CSR or integrated reporting.

GHG accounting is a misnomer from the pure accounting perspective as it defines GHGs as an asset despite to its non-availability beyond release to the ambient environment and the difficulties in ascertaining ownership that allow firms to exercise unusual control due to a lack of traditional property rights, which is mandatory for an asset to be defined as one. Still, GHG accounting allows the development of an emissions accounting system on the lines of a materials or inventory accounting system that can track physical flow. Accordingly, although the purchasing of emission rights can be handled by financial accounting, its non-fungible nature is an issue, not to mention the disconnect between financial and GHG accounting as well as the contracting challenges of emission rights and emission trading. This is where the coexistence of parallel streams of accounting would be of value: while the financial part of dealing with carbon-based transactions like emission trading, rights purchases, CDM projects, joint implementation (JI) projects and emission allowances can be handled by financial accounting, environmental accounting can handle the transactional aspects of emissions.

A key debate in here is who should bear the load of the bulk of the emissions. Currently, GHG accounting has used producer-based emission norms in some cases (Scope 1) and consumption-based norms in other cases (Scope 2 and Scope 3). Scope 2 and Scope 3 relate to the consumption of electricity and other services (after buying them from producers) that contribute to the production, manufacturing and conversion processes. As a result, responsibility for the emissions of power producers is distributed to the consumers based on the electricity consumed from non-renewable resources. In the absence of an actual profile of power producers who are supplying to any grid, emission loading on consumers would be incorrect, as power suppliers could be using different fuel sources (fossil fuels, renewable resources, biofuels or hybrid production). For example, if a new power producer is contributing to 10 per cent of the unmet demand in the grid using renewable resources, it would be improving the carbon load by 10 per cent. However, if the same producer is creating surplus supply for a grid that is green to an extent, say 40 per cent, it is not going to improve the green credentials of the grid. As against production-based emission, which drives producers to reduce the emissions, consumption-based strategies transfer the load to the ultimate consumers and penalize them for bad choices. Other than the time delay that would occur in fixing the responsibility, the shift in responsibilities is a major concern in the case of emission related-liabilities if they get transferred to the end of the chain. This is a representation of economic gains for producers, whereas the losses are pushed out for society to bear. The last complication brings in the forestry, lumber and timber industries that deal with sequestered carbon (carbon sinks) and agro-based industries who would need to apply a different methodology based on changes related to LULUCF and forestry and other land use (FOLU) norms. For example, if an agro-industry is involved in producing tobacco, would the social cost of tobacco consumption be comparable to the environmental benefits of saving carbon by having tobacco plantations?

At the supranational level, the inability of our leaders to go beyond the realm of nation state and further the economic interests of their regions hiding behind the populist agenda of economic sovereignty is quite visible. In spite of signing global agreements, most of the leaders from developed as well as developing countries are not interested in mandatory reductions of emission levels, and instead would prefer the voluntary reduction of emission levels to persist. Emissions being historical as well, their cumulative ownership is a difficult negotiation. Even if developing countries could force the developed countries to own their share of the historical load, which is also difficult due to the lack of acceptable accounting measures, mitigation efforts would still have to be enforced at local levels. Beyond these, the disagreement between different blocks of nations is reflective of our desire towards materialism without being accountable for it. This lack of accountability in our national lives is reflective of our inability to enforce accountability as a collective bargaining process or of a lack of knowledge to accomplish it. This is where accounting gets divorced from the underlying accountability it is supposed to be built on. Accordingly, it is not clear whether carbon accounting and related methodologies would anyway be effective in contributing towards societal improvement beyond the legal or regulatory framework and be able to lay down the responsibilities of both sides in black and white.

INTEGRATED WASTE AND EMISSION FLOWS

This extension proposes a theoretical construct to develop a unified flow of waste and emissions by integrating GHG accounting based on the materials and energy flow. As explained earlier, MFCA is a close variant of process costing, and depends on mass balance to calculate waste, it would let the GHGs from operations go unnoticed. Since the build-up of GHG inventory and its sequestration are the direct results of organizational activities, it would be beneficial for organizations to consider emissions as an integral part of their environmental decision-making framework, instead of resorting to parallel and/or ad hoc mechanisms. On the contrary, an independent GHG accounting system could compute GHGs at gross levels and allocate these to individual processes, irrespective of the actual emissions of processes, similar to the financial allocation process for overheads. This might lead to incorrect emission profiling for the processes and identification of hotspots.

One of the ways to achieve this could be to use the existing methodologies of EMA to capture and account for the GHGs arising due to organizational activities. To improve the visibility of emissions, the energy flow could be used to calculate and inventorize the GHGs that are created or sequestered through business activities. Although energy flow is not studied independent of the materials flow per MFCA, the complementary strengths of both these methods show promise beyond the isolated mapping of waste streams. In the absence of an integrated framework, however, firms would have to implement separate systems to capture materials waste, energy flow and emissions. Instead, a better way would be to create a GHG metric that is causally linked to energy consumption or any other factor that leads to the generation of environmental aspects as part of the underlying process. This would help in identifying the hot spots better and improve the accuracy of emission profiles. Complete details of the construct and a sample case can be referred from Debnath (2014).

Formulation

If a and b are the input materials participating in the process that results in the finished goods c and d, and E is the resulting difference in mass, then as per MFCA, the waste created in the process is:

$$(E) = (a + b) - (c + d) \qquad (15.1)$$

Accordingly, the waste stream would be valued (per MFCA) as:

$$E \text{ (in \$)} = E(Q) \times [(a(Q) \times \$a + b(Q) \times \$b + CC] / [a(Q) + b(Q)] \qquad (15.2)$$

where (Q) = equivalent physical quantities,
 $ = rates per physical unit,
and CC (conversion costs) = $\Sigma\ RQ_i \times \$_i$, where RQ_i = Quantity of ith resource

The uniqueness of MFCA is not in the summation of the elements but in following the iterative process to scale up the costs through every process and load the value on the outputs proportionately. On the other hand, as per GHG accounting, GHG produced during the process would be:

$$\text{GHG (in tCO}_2\text{e)} = \text{GHG }(a+b) + \text{GHG (CC)} \qquad (15.3)$$

Further, GHG (CC) = GHG (energy)
 + GHG (other resources) (15.4)

It is obvious that both the frameworks depend on the materials and energy used during the conversion processes. Accordingly, these could be integrated by leveraging the commonality of the processes. The proposed unification can be formulated through the following steps:

Step 1

Consider a manufacturing process comprised of multiple stages of conversion, sequenced as 1, 2, 3, ... n. MFCA could generate data on fuel consumption at every stage of the conversion, and the associated costs. Using the standard calorific value of the fuel (E_1, E_2, E_3,... E_n), the equivalent energy content (EF_a, EF_b,... EF_x) consumed during every process can be calculated.

Step 2

The total emissions for the period would be the sum total of the respective emission quantities (G_1, G_2, G_3,... G_n), calculated on the basis of the emission factors corresponding to the individual fuel types. If G_{total} represents the total emissions for the cumulative energy consumption during the period 't', it can be calculated as:

$$G_{total} = \Sigma\Sigma\ E_{nf} \times EF_f \qquad \text{during time period 't'} \qquad (15.5)$$

where E_{nf} = Equivalent energy input summed over 'n' processes for fuel type 'f',
and EF_f = Emission factor for specific GHGs corresponding to fuel type 'f'.

To account for the sources and sinks of GHG emissions, the equivalent energy value can be added or reduced depending on the direction of use. The modified expression can be represented as:

$$G_{total} = \Sigma\Sigma\ s_f \times E_{nf} \times EF_f \qquad \text{during time period 't'} \qquad (15.6)$$

where s_f = source or sink (represented as +1 or −1 or vice versa)

Depending on the creation or sequestration of GHGs in a process, the emission quantities can be recorded as positive (debits) or negative (credits) modifications to the emissions (GHG) stock account. This process covers Scope 1 and Scope 2 emissions.

Step 3

To build Scope 3 emissions into this construct, a GHG profile associated with support (non-production) activities has to be created. The

Scope 3 emissions can be loaded to the production activities based on direct calculation of GHGs.

In this experimental construct, MFCA plays the role of providing the basis of causal linkages to individual processes within the value-chain functions and associating GHG inventories to their sources. In contrast to manufacturing industries, service industries could use energy profiles (instead of mass balance) to develop a similar construct with equal ease.

Validation of the experimental construct using data from FPP

Table 15.1 shows the generation of the GHG profile based on the consumption of energy and transportation of materials at FPP (case 1 from Chapter 13). Here the transportation of raw materials and finished goods is converted into product travel (t-km), before applying conversion norms to derive the GHG stock produced and measured in equivalent tonnes of carbon dioxide (tCO_2e). The GHG profile in Table 15.1 is based on the emission factors selected for the conversion process as per the United Nations Framework Convention on Climate Change (UNFCCC) Project Reference #1497 (Fresenius Kabi India Private Limited 2009). The emission factor for electricity is as per the Central Electricity Authority (2011) and the transport load is as per the emission norms of Indian road transport for trucks and lorries (heavy-duty diesel engine with capacity > 3.5 tonne/gross vehicle weight: Ramachandra and Shwetmala 2009) for a product-travel aggregate of 186,000 km inwards (for 461 consignments of raw materials) and 1,920 km outwards (for 96 consignments of local dispatch with an assumed average distance of 20 km).

Next, the limitations of the traditional GHG framework, which works with aggregate levels of resource consumption, is displayed. In the absence of GHG profiles for individual processes, organizations may choose to allocate the overall GHG quantities to the individual processes by using a suitable allocation driver, as worked out below. Assuming an annual GHG load of, 1048 $mTCO_2e$ (due to energy and product travel only), suitable process drivers could be used to perform a step-down allocation of emissions over five selected processes

Table 15.1 Step-down allocation of aggregate GHG to different processes (in absence of GHG profiles)

Particulars[a]	Total	P1	P2	P3	P4	P5
Total unallocated emissions	1,048,752	–	–	–	–	–
Step 1 allocation using energy consumption (924,343 kWh)	(986,419)	–	100,201	599,252	286,966	–
Step 2 allocation using transported quantities (9,800 tonnes)	(62,333)	31,802	–	–	–	30,531
Allocated emissions	–	31,802	100,201	599,252	286,966	30,531
Actual emissions	1,048,752	11,703	92,017	630,873	263,529	50,630
Under/(Over) allocation (Actual – Allocated)		–20,099	–8,184	+31,621	–23,437	+20,099

Source: Author.

[a] All figures for emissions are in tCO$_2$e, unless otherwise specified.

(Table 15.1). The first-level allocation of GHG is based on the electricity consumption and the second level is based on the application of transportation tonnage (product travel) to the residual (unallocated) amount.

In the second instance, GHG allocated to the same five processes from the above exercise is compared with real GHG profiles that have been derived by the application of an extended framework (Table 15.2), which reflects the inaccuracies of driver-based allocation.

You can see in Table 15.1 that driver-based allocation can result in over- or under-allocation of emissions, resulting in incorrect identification of GHG hotspots, and could lead to disastrous results while undertaking corrective action. The role of environmental accounting as part of this construct is to develop data on the environmental aspects while maintaining the causal linkage with the energy consumption recorded as part of the cost accounting records. Data from environmental accounting can directly help the firms identify the GHG hotspots. In that sense, the traditional cost-accounting framework converges with environmental accounting at the transactional level to develop an information base that could provide better visibility into resource usage and environmental aspects, leveraging the temporal profile of activities that are not a result of quantitative evaluation.

Emission accounting due to transportation

Using product dispatch data from FPP, we can see the change in the GHG patterns, if the perspective is shifted from producer-based liability to consumer-based liability. For this purpose, product miles generated by dispatch of finished goods from FPP over a period is being detailed here. Based on 20 km average domestic travel, the converted product-km works out to be nearly 96,000 km-t. Using the standard emission norms for trucks and lorries in India (GHGs 2,417.95 g km^{-1}; SO$_x$ 1.42 g km^{-1} and PM 0.2 g km^{-1}: Ramachandra and Shwetmala, 2009) for 345 containers, the total emissions for the distance travelled work out to be around 16.68 tCO$_2$e GHGs, 10 kg SO$_x$ and 1.3 kg PM. Along similar lines, the export of 922 tonnes of

Table 15.2 GHG profiling of organizational processes (using the proposed framework)

Processes	Raw materials transport (Scope 3)		Conversion (Scope 2)	Energy consumed (Scope 1 and Scope 2)		Finished goods transport (Scope 3)		Emissions profile
	(t-km)	(tCO$_2$e)	(tCO$_2$e)	(kWh)	(tCO$_2$e)	(t-km)	(tCO$_2$e)	(tCO$_2$e)
P1	22,190	11,703						11,703
P2				93,895	92,017			92,017
P3			80,562	561,541	550,311			630,873
P4				268,907	263,529			263,529
P5						96,000	50,630	50,630
Total	22,190	11,703	80,562	924,343	905,857	96,000	50,630	**1,048,752**

Source: Author.

pulp equivalent to different countries totals 10.286 million km-t miles over 57 consignments, using the average consumption norms of a container ship for volume goods, at 0.0089 kg per tonne-kilometre, or tkm (Schmied and Knörr 2012), works out to be 92.403 tCO$_2$e for the distance travelled (Tables 15.3 and 15.4).

This leads to the question of how product miles can be reduced by the transportation industry through route optimization or by balancing product miles with the corresponding financial gains. Another option is to find technological solutions and other sources of fuel that can reduce the product miles. However, a third option is the one that can achieve the best results, but is the hardest: To meet our demands through a proximate supply. This would need societies to change their prefernces, policies and mindset so as to source and procure

Table 15.3 *Domestic dispatch (in tonnes)*

Product	May	Jun	Jul	Aug	Sep	Oct	Nov	Dec	Total
Amla pulp			8				36	26	70
Alphonso Mango Pulp	556	834	200	165	134	256	132	61	2,337
Banana Pulp + concentrate	7	80	126	8	170	76	40		507
Mango juice			0		4	58			62
New Mango Pulp					3	18			22
Old Kesar Mango Pulp		208							208
Papaya juice			4	48					52
Mango pulp	69		65	22	441	294			892
Guava pulp				22	35	22			79
White Guava Pulp			0	5	230	315	17		568
Grand total	632	1,123	408	496	1,103	742	207	88	4,798

Source: Author.

Table 15.4 Export dispatch (in tonnes)

Destination	Nautical miles	May	Jun	Jul	Aug	Sep	Nov	Dec	Total
Busan	4,938	18.24		18.24					36.48
Fos-sur-Mer	4,558			17.08	50.4				67.48
Jakarta	2,708	68.8		0.25					69.05
Los Angeles	10,104				17.6	17.6		68.8	104
Miami	8,795					17.6			17.6
Rotterdam	6,320				37.85				37.85
Singapore	2,435				35.787				35.787
Toronto	8,430		17.2	17.6	17.2		17.2	34.4	103.6
Vancouver	9,513			17.2					17.2
Yokohama	5,327	64.64	77.68	138.16	92.8	45.36	15.12		433.76
Grand total		151.68	94.88	208.53	251.637	80.56	32.32	103.2	922.807

Source: Author.

products locally (say, small-scale cooperative zones) and live with limited demands, a much larger sacrifice!

CARBON ACCOUNTING USING ENVIRONMENTAL ACCOUNTING CONSTRUCT

This extension is based on the need to institute carbon accounting in a way that has relevance to the carbon trading in emission markets. Annex B of the Kyoto Protocol lists Annex I countries that need to enter into binding agreements to meet GHG reduction targets within their individual commitment periods. Governments of the respective countries are free to fix emission ceilings for their emission-intensive industries. These binding emission targets are a way to create an artificial shortage of carbon in the market, to offer economic incentives to the industries to reduce overall GHG exposure and for the respective governments to honour their share of commitments under the Protocol. Countries are free to establish compliance markets, to allow the trading of emission allowances in the form of allowable accounting units (AAUs) that are granted to the industries and to administer these through suitable regulatory mechanisms. Organizations are allotted AAUs with an absolute threshold, beyond which they are not permitted to generate emissions. Emissions generated over and above the agreed levels may be penalized through compulsory purchase of emission units to cover such shortfalls at a penalty rate (Bebbington and Larrinaga-González 2008).

Following the emission regulation norms, firms from these countries are expected to manage their GHG exposure and keep it below the allowed ceiling to avoid penalties. They can reduce their obligations by purchasing additional AAUs from the compliance market (if allowed and if available) and/or acquire CER rights from the voluntary emission markets. Also, firms are free to generate CERs by participating in registered CDM projects with firms from non-Annex I countries or as JIs with firms from other Annex I countries. To make financial gains from lowered emissions, firms might as well sell off excess AAUs in a cap-and-trade market or trade CERs in the voluntary market, subject to legal norms (UNFCCC 2008). While individual countries can have multiple such instruments and market mechanisms, the present discussion analyses sample transactions from such interactions without assigning any contextual reference.

Example:

Assuming the total emission of an organization ABC (under a capped regime) is α tCO$_2$e at an operating level of X and that ABC is planning to operate at level Y, the allowed emission level would be:

$$\text{Emission level allowed (legal limit)} = A \text{ tCO}_2\text{e}$$

where total potential GHG at Y level of business activity

$$= \alpha \times (Y/X) = B \text{ tCO}_2\text{e} \qquad (15.7)$$

$$\text{Difference} = A \sim B = C \text{ tCO}_2\text{e} \qquad (15.8)$$

If C>0, then ABC is well within its capacity to carry on its activity. Still, ABC can further reduce emissions by participating in CDM or JI projects, sell excess AAUs and/or acquire CERs from the relevant emissions market. On the contrary, if C<0, this leads to a shortfall in the projected emissions level that would have to be covered by augmenting emission rights through a CDM/JI project and/or purchase of additional credits, AAUs and/or carbon credits from the emissions market to cover the shortfall. The ensuing discussion explains that carbon accounting is not just limited to financial accounting. It needs an additional accounting construct to collect, hold and transfer CERs, AAUs and actual emissions (in equivalent Kyoto units). Conventional accounting constructs are struggling to provide suitable mechanisms to handle the situation (Bebbington and Larrinaga-González 2008). Environmental accounting proves to be helpful here by separating the accounting treatment of the transactions based on the area of interest. To illustrate the point, five types of transactions from the carbon trade cycle are covered in Table 15.5.

Business activities and environmental stewardship should make a windfall of such situations and leave the organization with a situation where (B) − (C)<(A). As can be seen, the financial viewpoint does

Table 15.5 Carbon accounting, leveraging environmental accounting

Type of transaction	Financial accounting	Environmental accounting
1. Accounting the initial AAUs (as assigned limits) gives the emission ceiling of an organization. This should be accounted for in the financial and environmental dimensions, both.	These may get accounted as grants (IAS 20) and intangibles (IAS 37) on allocation. No specific directive is available from GAAP, yet (Machado, Lima and Filho 2011). Care needs to be taken so as not book the notional profit/loss due to the difference in valuation at the end of the commitment period.	AAUs are accounted as budgeted (allowable) emissions in Kyoto units against contra of environmental liability (budgeted).
2. Purchase of additional CERs/AAUs from the emissions market to cover the shortfall of emission levels at the prevailing rate, for business reasons.	Funds are to be depleted to acquire these. The Kyoto units are to be added to intangible (AAUs) or tangible assets (CERs), as the case may be.	Kyoto units covered within CERs would be credited to the emission account and AAUs would increase the environmental liability (budgeted).
3. Sell available CERs earned through CDM/JI projects in the voluntary market and/or surplus AAUs released in the compliance market. (So opposite of the previous transaction.)	Funds are received due to the sale. The corresponding asset pool is depleted by the sale value of relinquished/transferred items and any profit/loss due to the price differential of the assets is booked. The issue of generating a notional profit/loss due to the sale of the initially allocated AAUs remains.	Debit emissions account by Kyoto units sold through CERs. For AAUs, debit the environmental liability (budgeted) a/c

(Continued)

Table 15.5 (Continued)

Type of transaction	Financial accounting	Environmental accounting
4. Earn CERs through a CDM investment. This type of transaction is possible due to the earning of emissions credit due to participation in CDM/JI projects.	Financial accounting records the acquisition of such value-based assets from CIP (Construction-in-progress projects; the underlying CDM/JI project). The Kyoto units gained through the CERs get averaged due to total investment (?).	CERs earned would reduce the emissions account by the corresponding number of Kyoto units.
5. Kyoto units of emissions generated due to production or other activities: This entry would detail the extent of consumption of Kyoto units.	Not applicable.	Debit the emission account from Kyoto units and credit the environmental liability (offset) account.
6. Period-end impact.	a. P&L (Profit and Loss Statement) flow = (3) Reasoning: Profit/loss due to sale of AAUs/CERs. b. Balance-sheet flow = (1) + (2) − (3) + (4). Reasoning: Remaining Kyoto units as assets.	a. Total AAUs at the end of the period: $(A) = (1) + (2)_{AAU} - (3)_{AAU}$ b. Total actual emission: $(B) = (5)$ c. Total CERs: $(C) = (4) + (2)_{CER} - (3)_{CER}$.

Source: Author.

not offer visibility into the consumption pattern of AAUs due to the emissions generated from business activities or the signal coverage required to avoid a penalty for excess emissions. Moreover, AAUs being legal limits and intangible in nature, should not be mixed up with or set off against real emissions, because unlike allowances, real emissions are physical assets generated by the firm. Currently, the accounting process is broken and being carried out in an ad hoc manner, not to mention the unavailability of any other accounting framework as well as the lack of standard accounting constructs from the IFRS. The accounting of emissions using environmental accounting would maintain the overall transaction and link with the financial transactions, resulting in a vastly improved information flow. This not only contributes to financial accounting but also improves the decision-making process by increasing the visibility of the carbon exposure and environmental performance of the firm.

Environmental Accounting and Managerial Implications II

Other Advances

16

INTRODUCTION

In this chapter, we continue the discussion on how environmental accounting would improve information needs of management and help firms in their tryst to improve environmental performance. To this end, first we experiment with an extension which is proposed as aggregation of aspects at micro-level (firms) and to generate meso-level (industry) indicators that can help policy makers to review the impacts of economic development at regional levels, followed by reviewing latest developments within framing integrated reporting. Last part of this discussion is to explore practical aspects of developing a green information flow where integrated view on environmental aspects could be automated as a part of green Information System that would extend information base and help firms and their management develop business excellence while reducing impacts on environment.

Environmental performance at the industry or sector level: Developing a meso-level view

By aggregating the individual performance of firms, we can build the industrywide or sector-level picture of performance that this extension proposes. The WBCSD (2000a) has long been a proponent of eco-efficiency and GRI (2006) reporting has helped organizations to use eco-efficiency to improve the organizational footprint. Also, the IFAC (2005) advocated using cross-cutting EPIs that could be linked to economic performance. Rao et al. (2009) developed corporate environmental indicators for the SME segment in the Philippines by combining multiple frameworks such as environmental information system (EIS), sustainability metrics indicator, ISO 14031 and the environment standards proposed by European Environment Agency (EEA). In contrast, Dawkins and Fraas (2011) correlated the quantum of voluntary environment disclosures by companies in Japan with their environmental performance using KLD performance rankings.

However, experts are divided on the subject of ascertaining the exact relationship between disclosure levels and environmental performance. For example, using S&P 500 members[1] with KLD performance rankings as the benchmark for environmental performance, Huang and Kung (2010) found that the level of voluntary environmental disclosure had a U-shaped relationship with environmental performance, indicating the worst as well as the best environmental performers would tend to have a higher voluntary disclosure levels. This corresponds to the legitimacy theory where by disclosing environmental performance (voluntarily), the best performers develop this as a differentiator which improves their image, whereas the worst performers avoid the likelihood of being perceiving as a wilful polluter or an unconcerned citizen. Moreover, corporate environmental performance and systemic perspectives on it could connect the overlapping areas of EMSs, eco-control and financial systems. In one of the studies involving the UK water industry, the non-comparability of environmental data reported by the member organizations across organizations and time periods was highlighted. The study reflected the need for standardized EPIs that could be developed and applied across industries to generate comparability and consistency and to make these reports meaningful (Hopkinson, Sammut and Whittaker 1999).

Ienciu (2009) discovered theoretical linkages between corporate environmental performance and other aspects of environmental concerns, and proposed a model to derive the environmental profit (defined later). To improve environmental impact and develop a

---

[1] The S&P 500 stock market index, maintained by S&P Dow Jones Indices, comprises 505 common stocks issued by 500 large-cap companies covers about 80 percent of the American equity market by capitalization.

win–win strategy, he felt that the organizations needed to improve their technology, attention to regulatory systems, visibility, WTP for alternative materials and processes, and benchmarking practices, so that both extra costs and resulting profits would be generated by improved, environmentally conscious products. Labuschagne, Brent and van Erck (2005) analysed different sustainability frameworks such as the GRI, the United Nations Commission on Sustainable Development (UNCSD), the sustainability metrics from the Institution of Chemical Engineers and the Wuppertal sustainability indicators to reveal the lack of real-time support in managements' decision-making due to the lower level of synergy between indicators across frameworks. They proposed the use of institutional aspects to cover institutional sustainability, business strategies, social investments and sustainable business initiatives, and modelled the cross-linking of micro-level (institutional) events with macro-level parameters. In another study from Japan, Xie and Hayase (2007) evaluated multiple third-party environmental performance-evaluation standards that are used by private parties to rank organizations by their corporate environmental performances. However, the evaluations were found to be limited and arbitrary in nature and lacked a theoretical premise.

This led to proposing standardized, theory-based, aggregated measures by indexing environmental performances from a specific base year, which could be used to evaluate relative improvement in environmental intensity in subsequent years. There are practical limitations in using eco-efficiency as a measure and to compare and improve efforts towards sustainability, however, which have been discussed in Chapter 8. For this purpose, the inverse of eco-efficiency—that is, eco-intensity—was employed to develop eco-intensity change index (EICI), which could be used as a construct to aggregate the comparable performances of firms from an industry in a given year. This construct can cover multi-year analyses of sectors, besides offering a platform on which to compare the environmental performance of individual sectors with their respective financial performances in aggregate, and to derive first-degree information on the incremental environmental gains and losses. This corresponds to the assumption that using eco-intensity based indicators enables us to link micro-level and meso-level economic activities by using suitable aggregation methods. For example, using eco-intensity, a temporal comparison within and across sector levels can be generated and benchmarked. Meso levels can in turn fold to macro-level performance of a region and can signal changes in the resources and environmental conditions of the region. Needless to mention, the progress of green accounting at the national level has not been able to integrate micro-level performance (as a part of the SEEA or similar other frameworks) due to the causal disconnect between macro-level goals and micro-level performances. This development might help in the formulation of pro-environmental policies by the regulatory authorities, even though due to the inherent nature of eco-intensity as a ratio, variations across industry sectors cannot always be attributed to industry performance alone. Here, the discussion is limited to abstract the construct:

Step 1

The absolute values of the indicators reported by the participating firms in a year within the same sector and using the same units are summed up. The aggregate values are divided by the selected financial aggregate arrived at in similar manner. In other words, the ratio of environmental performance for indicator 'x' for sector 'j' and year t would be $EnI(x)_{tj}$ and the selected financial performance $FinBase_{tj}$ would give the environmental intensity for the indicator 'x' as $EI(x)_{tj}$:

$$EI(x)_{tj} = EnI(x)_{tj}/FinBase_{tj} \qquad (16.1)$$

Both the variables could be rationalized to keep an equal number of observations on both the sides, dropping missing observations that has only one variable reported.

Step 2

Derive the ratio of environmental intensity over the base year and relative change as compared to the base year. So, if $EI(x)_j$ for year t+1, t+2,... is divided by EI(x)$_j$, 't' being the base year, the EICI would be derived for successive years. The comparative change across the years would indicate the temporal movement of the environmental indicators for industry sector 'j' with respect to the fixed base year 't'.

$$\text{EICI (x)}_{j(t+1)/t} = \text{EI(x)}_{(t+1)j}/\text{EI(x)}_{tj}, \qquad (16.2)$$

$$\text{EICI (x)}_{j(t+2)/t} = \text{EI(x)}_{(t+2)j}/\text{EI(x)}_{tj}, \quad \text{and so on} \quad (16.3)$$

Step 3

If EICI (x) is greater than one in any year, it would mean increased environmental impact per unit of financial performance for the sector for the specific indicator, signifying bad performance (and vice versa, if less than one). Comparative values across years would indicate the resource intensity of the sector and its direction.

The relevant details to develop the construct can be found in Debnath, Bose and Dhalla (2014), where the construct has been verified using three-year data from 16 Indian organizations (environmental performance reported to GRI during the period 2007–2010).

In this exercise, the role of environmental accounting is relevant in reflecting the reporting of authenticated data (if institutionalized) and helping policy and decision makers with new information, which could also be assured per audit guidelines. Accordingly, environmental accounting should not only help firms improve the accounting of environmental aspects, but also enable objective reporting of their activities and enables them to measure performance against the benchmarked values of the sector to work towards improving it. However, grossly overgeneralizing the findings could be a mistake, as the results would depend on the number of selected firms and other comparative studies. Also, the effort should refrain from standardizing EICI, as the intention should not be to rank firms or sectors but to study the suitability of eco-intensity for sector-level analyses, where standardization can be achieved by ranking performance across sectors to help policy makers with better information on the sectoral profiles. Sectoral differences inherent to the data could have some bearing on how the data is classified. Jollands, Lermit and Patterson (2003) have demonstrated the possibility of using a similar aggregation by using panel data for New Zealand across multiple periods and industries. With reference to any other country or region, this would need the development of a data bank to capture the environmental performance of industries and for policy makers to work towards slowing down the increasing trend of environmental litigations (industrialization versus pro-environment activity fronts), which are consuming valuable resources in debating economic growth versus environmental protectionism.

INTEGRATED REPORTING AND FRAMEWORK(S)

This subsection is a reflection on and an overview of the sunrise practice of integrated reporting and its intent to reflect sustainability. Reporting of organizational concerns and behaviour, and their connect with societal demands have been theorized in stakeholders' theory (although equating them to management responsibility), which businesses are expected to be participating in and contributing to. Primarily, the participation of industries is viewed as necessary to improve the collective response towards climate change and sustainability, intergenerational equity, equitable distribution of wealth, reduced consumption of materials and energies, and other concerns of society. The demand for an improved form of reporting has generally been related to the insufficiency of disseminated contents. Gaps in the disseminated information as against expectations reflect (a) a disconnect between business firms and their participation in addressing the impending concerns of society and (b) a reality check for the actions being taken to address some of these concerns, which leads to a crisis in public confidence. Integrated reporting brings a new perspective that is believed to bridge this gap (Beck, Dumay and Frost 2017). The academic debate around integrated reporting includes aspects like the convergence of reporting in literature and the regulatory adoption of some of these in select geographies.

Financial reporting and its dissemination has been a subject of intense study for long but has been found to be of limited use in communicating organizational efforts towards social and environmental concerns, contrary to what the proponents of SEAR expected. It is worth noting that SEAR was proposed mainly by the academic fraternity with the expectation that the voluntary reporting by the firms would include details of organizational efforts towards sustainability. That was a tough ask, but the hopes were that the firms would move beyond a position that essentially suggested 'trust me', without bringing in regulatory and legal pressure. Although the firms fulfilled a part of the expectation and voluntary reporting gathered some momentum in one form or another, organizational stakeholders, academicians and societal stakeholders did not always find the disclosures comprehensive (covered in Chapter 7).

Integrated reporting is a very recent development and is waiting on the scholarship for an apt definition. So far, it has been used to mean different reporting frames—like sustainability reporting, social and environmental reporting, integrated reporting by the International Integrated Reporting Committee (IIRC), or even integration of reports. From the perspective of usability, although it can be a pragmatic solution to integrate a multiplicity of reporting frameworks (voluntary as well as regulatory ones) that are currently in vogue and/or are being promoted by different agencies (e.g., the UN Global Compact, GRI, IIRC, SEAR, the Sustainability Accounting Standards Board, or SASB, the GHG Protocol and CSR) other than the financial reporting, the challenging aspects of integrating them have academic, political and institutional overlaps that have kept things open and uncertain.

The genesis of a multiplicity of these frameworks and their operationalization reflects our belief in the democratic process and in polylogical debate in anything that concerns institutionalization, but that is out of the scope of this text. From the above, reporting frameworks like those from the GRI, the GHG Protocol, IIRC and SASB have been promoted by institutes that operate through business representation, or in other words, mainly from the practice side. Accordingly, these frameworks are promulgated and evangelized mainly by the practitioners' bodies to enable business firms to take up voluntary initiatives. These frameworks have been passed through business case framing where businesses and practitioners' bodies have been invited to be a part of the framework-setting process. While this could be a concern from academic perspective, as the frameworks lack deep academic thinking and rigour, it is expected that the participation of firms over time would drive the improvement of standards and quality of disclosures. Though we will get into more detail a little later, suffice it to say these frameworks are different from the financial reporting needed by regulatory agencies and led to the practice of multiple streams of reporting in different markets. In some cases, integrated reporting could have been tried, but that was before a new set of standards was established by the IIRC, where integrated reporting has been promoted as a new framework.

This leads to two different questions, the first concerning the legitimacy of the frameworks or their governing institutions and the second asking what we are integrating.

The legitimacy of these frameworks is a challenging topic, as most of them are not a part of a regulatory mechanism, nor endorsed by a regulatory mechanism or bound by legal provisions, in contrast to the financial reports that firms are legally bound to produce as per the provisions of the companies laws or similar statutes in different countries. Moreover, the setters of the standards are not involved in defining the reporting framework or in deciding the scope, coverage or contents of reporting (Dumitru and Guşe 2017). The authors' employed legitimacy framework to analyse different stakeholders' involvement in IIRC and SASB to hypothesize how different countries would react to it. Therefore, obtaining the involvement of the academic community is challenging. This being the case, it should not come as a surprise if we find the disclosures are of limited value and do not reveal what form of sustainability firms are involved with, in terms of economic, social and environmental paradigms.

There has been a good number of articles from accounting practitioners' bodies like the American Institute of Certified Public

Accountants (AICPA) and the Big Four audit firms (Deloitte, Ernst & Young, KPMG and PricewaterhouseCoopers), offering commentaries on why and how these reports are good for stakeholders and markets. However, critical theorists have raised concerns that voluntary reporting has not improved how the firms operate or how they work towards improving the current state of affairs beyond economics; rather, it has helped firms in improving their market position by being a part of a voluntary reporting bandwagon. What is missing here is a prescriptive framework that is or would be jointly promoted by the accounting practitioners and by academia, with collaborative effort put in to cover the distance between the two groups. In my view, the lack of regulatory compliance is challenging, but has been put forth by the advocates of voluntary frameworks to help firms adapt to it, and advance efforts in their own way, to help the society achieve beyond what regulations can achieve (James 2015). A noble thought, I would say, and would have worked better in a society that is ethically driven. Arguments in favour of improved subscription to voluntary reporting need to be associated with the costs of reporting, interpretive risks and legality of what is being reported. But that is altogether a different debate, detailed later.

This brings us to the other question, asking *what* is getting integrated. Integrated reporting can be deconstructed to mean the integration of multiple reports that an organization disseminates as a part of compliance or otherwise. In that sense, integrated reporting is more about combining reporting artefacts and contents. Obviously, it would help to provide a narrative that improves the coherence and materiality of the reporting, rather than the parts being read in isolation or some areas remaining under-reported or simply ignored. As with any practice area, Integrated Reporting is yet to connect to reflect firm's business or operating model and its sustainability drive, which brings in descriptive (whatever is the situation why this is so), prescriptive (what should happen versus what has happened), and normative (what ought to have happened versus what has happened) concerns and evaluate real value of a firm. The integrated form of reporting should integrate commonly reported information under a single umbrella, one would hope, including financial as well as non-financial information, such that the users of the information can find not only quantitative data, but also accompanying explanations detailing:

- how the firm has been performing,
- how it is planning to continue the journey ahead,
- what business changes it is planning to bring in,
- actions it has taken to reduce the consumption of natural resources
- what it means for the organizational supply chain to accommodate sustainability challenges,
- what business model the firm is considering to improve its business,
- how that would contribute to the region where the firm would be operating, and so forth.

While not trying to be overtly prescriptive, it is more about how the firm is creating shared value and embedding sustainability, something it is expected to be aiming for in the near future. In contrast, the IIRC has come up with a framework for how a firm and its business model are creating value from six types of capital—financial, manufactured, human, intellectual, social and natural capital. Moreover, companies have started using the narrative of ethical positives to improve legitimation while adhering to the older reporting practices, rather than genuinely examining their impacts on ecology and the environment (Haji and Hossain 2016). So the integration is not about the process or to bring in the deep-green ecological perspective that one would expect to improve everyday processes and work lives that inspire value creation process, and instead boils down to narrate a story that mimics unification of multiple narratives. But what is the overall paradigm within which this is taking place?

Before we explore that, let us consider the cost of reporting.

The cost of reporting is associated with managing and tracking data and information (what is being reported), keeping track of changes in the standards and retrofitting previously reported data, upkeep of information systems, training and maintenance of staff and associated expenses, developing a mechanism to keep a track of reactions (mostly the adverse ones) and benchmarking costs. These costs, to a traditional mind, would need to be offset by, say, changes in the market valuation of the firm or its perceived position against the peers. On the other hand, interpretive issues are associated with lack of clarity in applying the reporting norms to a business case and providing justifications or narratives for changing positions or challenging situations. Similar concerns can also arise after dissemination, where the provided information does not offer a seamless narrative that it was supposed to in the first place. Also equally challenging is determining how firms and their management can be held responsible for the contents of the report and to what extent the law could enforce this. This is a typical case of incomplete contract, where firms and their management could become susceptible to what has been said voluntarily. This could potentially become a regressive situation for firms to deal with, which might improve with integrated reporting, provided the necessary convergence happens and, in all fairness, can be weighed in to allow innovativeness to firms with the constructs given or provide guiding principles for firms to follow, as in the case of financial accounting.

From the academic side, researchers view integrated reporting largely as a pro-market construct, which is also the case for other forms of reporting (more or less) and hardly challenges the prevailing paradigm (Brown and Dillard, 2014), despite two opposing attitudes:

- Academicians and scholars would like business firms to be more cognizant of the impending risks they are contributing to, but do not offer any prescriptive solutions that the firms can experiment with. In addition, there are only a few prescriptive models where the academia has collaborated with practitioners.
- On the other hand, professional accounting bodies have considerable influence over practitioners to experiment with new forms of information artefacts, processes and dissemination methods. Practitioners, being members of these organizations themselves, would be more than inclined to experiment with new developments so long as the framing fits the prevailing paradigm.

The last part of this discussion probes the issue of legitimacy. Existing management theories have used agency and stakeholders' theory to legitimize the need for management to experiment with the new frameworks, holding them responsible for the adoption and diffusion of new practices. However, this is different from the conventional paradigm of a top-down approach where the management's legitimacy is viewed as the starting point of all endeavours. Instead, integrated reporting could be a multidirectional approach, where good practices can be implemented and diffused in all directions and can involve everyone. Diffusion of a good practice does not need the management's exclusive go-ahead, so long as it is improving shared value and can be defined within the overall organizational boundary. Sustainability efforts (environmental proactiveness as well as contributing to social concerns) in organizations are effective so long as they get embedded within the processes and diffuse to become a part of every single activity within the organizational umbrella and outside. It may not always be one big change, but could be numerous small changes in one direction, and integrated reporting can bring this trend to light and relate it to an overall shift in the organizational behaviour.

GREENING THE INFORMATION NEEDS OF FIRMS

The accounting function of an organization is older than most others that define a modern organization. However, a modern organization is more complex and intricate, thanks to the information system that connects shop floor to boardroom. In a modern organization, different functions connect with each other through processes in which they participate and exchange information or share different contextual backgrounds, which are interpreted across functions depending on how data and information are used and consumed. Even accounting has transformed itself from being a manually driven exercise to using event- or activity-based triggers that translate physical activities into the equivalent accounting entry by using a certain pattern that is captured through a REA (resources, events, agents) model, which is a fundamental building block of an AIS (accounting information system). The ability of a REA model to comprehensively automate routine accounting events by abstracting physical processes is the underlying mechanism of automating accounting process and improving information flow in ERPs, which serve as the backbones of information systems in a modern enterprise.

Given the complexity of this background, what could be more challenging than to propose environmental accounting in isolation?

Accounting, being the repository of organizational history, can better develop its environmental angle for firms if it can connect with other functions, or if the greening of the firm can cover different business functions and processes, including the accounting and information systems as well. In modern firms, accounting is not done in isolation but depends on how different processes along the value chain are performed; hence environmental accounting would be better served if environmental aspects could be captured at source, right when a business transaction is captured in the reporting system. This will not only be closer to the regular accounting process, but will also enable environmental accounting to capture and translate environmental impacts. However, at the same time, this creates the need for 'operational fit' between the green business function and its information-carrying needs so that it can improve long-term sustainability of the firm. For example:

- Eliciting the need for suppliers to satisfy the supply criteria in green procurement, which are different than the existing supply criteria.
- In case of a supply-chain function, researchers have talked about extended information gain due to the greening function and improved decision-making, involving environmental considerations in different business scenarios.
- Sustainability considerations in manufacturing improves with the help of a manufacturing execution function (MES). By focusing on the resource and energy consumption function in different manufacturing processes, a redesigned MES is expected to capture and relay information to the sustainability information system of the firm, which would hopefully reduce resource consumption.

In theory, the sustainability outlook of business processes can be based on the transformative properties of information systems and a process-centric view of the supply chain. However, the interactions of processes and information within the domain of sustainability domain need to connect to other elements to understand (encode), capture (interface), process (decode) and disseminate (generate) the environmental orientation of processes. Moreover, if firms aim to implement EMS, explored further in the next chapter, they would generally target greening business activities instead of the accounting function per se, which in turn would need environmental accounting to generate environment-relevant information. This reflects a potential disconnect between the environmental considerations of business processes and their information systems, which need to be intertwined to support process-related changes, explored next.

Process-centric approach to understand environmental impacts

One of the common themes throughout the changing information needs of business functions upon greening one or more processes is the expectation for the information system to capture the impacts and fulfil related information needs, which unequivocally establishes that such a change would force the traditional information system to evolve. This is different from assuming that the information system is an intermediary that uses defined pathways to connect to people and processes, pointing to a rather static context for its operations. Even though improving the environmental sensitivity of individual business processes could lead to the development of process-specific information, it may also result in design considerations that are isolated and process-specific in nature, creating redundancies that might impact in achieving a holistic view of the needs of the firm for environmentally sensitive information. So, other than maintaining uniformity in the rule of engagements, process-specific advances (e.g., new emission calculation norms) would have to act as a conduit to connect processes (e.g., energy consumption involving different types of fuel). This would result in a green information system (or a green version of the traditional information system), with the ability to evaluate, process, forecast and report on the environmental aspects and impacts of organizational activities and to trace the environmental load of the processes over time. At the same time, green processes in firms would enable it to generate, translate and develop information that the management, decision makers and agents of change can use to steer firms towards improving the environmental sensitivity of their products and services, and in becoming more efficient, greener and sustainable.

However, very rarely have the researchers delved into how the interactions between green business functions and the information system would influence and reshape the information space to connect with the sustainability efforts of the firm. We would hardly get an insight into how the micro (functions) and macro (strategic choices) activities of a firm connect with sustainability themes and influence users to improve the current state of functions and processes after the greening efforts. With the information system maturing in the role of an enabler of environmentally enhanced information to meet the information needs of evolving business processes, it would have to engage with sustainability and its enactment in a way that would be dynamic, emergent and under-defined with respect to the nature of the firm's information needs as well as user expectations, which would in turn be subject to internal and external pressures. This is where process orientation leads to the development of a knowledge base of process elements and interactions. Greening the information system, if seen from an innovation perspective, can be studied using three approaches: diffusion of innovation (DOI), organizational innovativeness (OI) and the process theory of innovations (PT). The key question here is how green process-related innovations, solutions and tools connect and operate within the operating environment and shape the information space. To examine this, let's begin with DOI:

- Although DOI might fall short of examining the latter part of the query, yet it offers a much-needed theoretical lens to help in connecting the green information system to the operating environment and to delve in. DOI (Rogers 1983) is about the spread of innovations due to their adoption by adopters in the population within a given time period, where the unit of analysis is the innovation itself. Accordingly, much emphasis is placed on the phenomenon of adoption and the factors that impact its diffusion through the operating environment. Factors that influence the adoption process include variables such as the characteristics of adopters, innovation attributes, environmental characteristics and communication methods. Analysing these, we can stratify adopters along a time axis and predict the adoption and life-cycle profile of a process innovation. However, the DOI approach has been criticized for its inability to consider human, social and environmental contexts (Lyytinen and Damsgaard 2001) such that the black-box approach prevents any explanation of how innovations connect to the operating environment.

- In comparison, organizational innovation (OI) theory relates to uncovering the environmental likelihood of improving the rate of innovations within a firm. This approach includes cultural readiness (Zaltman, Duncan and Holbek 1973) and readiness to be creative (Kundu and Katz, 2003), viewpoints that characterize organizational capabilities to innovate, particularly related to diffusion. Although OI brings together a number of organizational variables that influence innovation, it has failed to establish the relative importance of any one variable versus the rest, resulting in a critical need for analysis and improvements (Baunsgaard and Clegg 2015).
- PT investigates the nature of the innovation process, including how and why innovations emerge, develop, grow and terminate (Wolfe, 1994). Arguably, PT research focuses on the conception, development, implementation and evaluation stages of an innovation, to connect the what (object) and the how (processes) of innovation. This resonates well with sustainability theory, where the relevant enactment to be 'sustainable' at the micro level is unpredictable, uncontrollable and emergent, and firms need to explore, experiment and play with possibilities, without worrying too much about accurately predicting the results, at least to begin with, and instead aiming to improve the processes and their organizational connect.

Here we can relate to Orlikowski's (1992) view that technology is not just an external object but a product of ongoing actions, designs and appropriations, a view that is relevant to how information systems might improve the information flow, to connect processes to environmental strategies and the firms' goals of firms. Other than the information on externalities, green processes would let information systems act as enablers of the organizational goals. The success of this plan would also depend on the ability and maturity of the users to use information imaginatively, which is an area open of discussion and depends on the users' motivations to go beyond a specific use of information. Still, newly emerging environmental view of activities helps us—at least theoretically—to expand to firm–environment interactions where business activities would have left their temporal prints. Corresponding to it is the two-dimensional frame of accounting, which when extended to environmental view, can leverage the temporal environmental prints of the corresponding business activities. This bi-directional correspondence (between business activity and the corresponding recording) is pragmatic and would lead to improve knowledge base including deciphering the complex physical phenomenon through focused views of accounting and in representing business transactions within green information system. This is where environmental accounting can engage with information systems, as I hypothesize the theoretical feasibility of extending the REA model to E-REA (environmental REA), where environmental impacts being captured along a viewpoint would also capture the environmental impact of underlying business events while connecting the two. I would invite researchers working with information systems to explore this further.

Environmental Management Systems and Greening Firms

INTRODUCTION

For firms to be environmentally savvy and to evolve sustainability-oriented business models, some form of institutionalized EMS and green information system is needed. Moreover, to become a progressive member of the community, firms from different industries have been adopting frameworks promoted by international agencies such as the International Organization for Standardization (ISO). This chapter will not detail the different forms of EMS available today in the certification market, although a review of these is essential and has been dealt with in the relevant texts; however, I seek to underline the need for EMS and for firms to adopt them as a part of their corporate character to articulate relevant information needs. The demand for new information in firms may result from implementing any EMS framework that contrasts information or differs from infromation generated by the existing accounting and information systems. This is not about the nature of information required for the organization to become environmentally relevant; it is more to engage in a discussion on green information needs against the contextual background of the framework that promotes it. The final point of the discussion is to examine whether the frameworks themselves are supportive of sustainability thinking in firms, and whether the new information needs could result in a permanent change in the environmental outlook of firms, referred to as 'morphogenetic changes' in the literature. This discussion seeks to relate the flow of information and how it connects with employees, processes, systems and operations with the firms' business objectives, with the EMS frameworks considered as the superstructure, and to provide for a much-needed contextual background.

EMS AND THE MULTIPLICITY OF ENVIRONMENTAL CARE FRAMEWORKS

EMS evolved as a response to the organizational need to address the growing demand from society and stakeholders to be responsible and diligent in how products, processes and systems connect with the environment along their life cycles. In generic terms, the establishment of an EMS relates to the need of the firms to institutionalize a framework that can cover environmental proactiveness in a systematic manner. Primarily situated within the context of quality frameworks covering plan–do–check–act (PDCA) cycle, these frameworks can be considered as the tail-end of the quality frameworks and are generally adopted by businesses only after successful adoption of quality frameworks. One oddity in the development of these frameworks is that EMS never had an organic origin (from requirements to implementation), in contrast to the quality frameworks that resulted from debates, discussions and refinements over the years, as the quality management systems (QMS) grew organically through successive stages of engagement, before expanding in response to and reflecting a cumulative view of business wants and evolving to deal with them. In contrast, EMS got institutionalized as a framework even before the relevant needs of businesses could evolve and mature to define beyond what quality and related frameworks have already offered. This resulted in solutions being offered before the needs were felt, a point counterproductive to its adoption.

If we refer to the literature to learn how quality became inherent to products and processes and how it has moved from being on the fringes of production to be at the centre of existential needs today, we can relate this to how quality-consciousness moved from being a choice to an inevitability in the design and life cycle of products, processes and services. The institutionalization of a framework standardized the communication of the underlying processes, while the firms' efforts to deliver quality in products and services to customers with transparency resulted in standardization where, along with the tools and methods, the processes and narratives were communicated through the adoption of a standardized model, signalling to stakeholders a certain level of quality-consciousness. However, when the institutions develop and promote environmental frameworks, it is surprising to discover that 'good quality' does not equate to 'good environmental quality' in their belief system. Could this be due to temporality of social conscience that delayed inclusion of social and environmental relevance or is it anthropocentricity that prevailed? Let's dig deeper.

Institutionalized frameworks are a way to impose control on business conduct or in general to bring a certain order to the certification or governance market, to channelize service demands from a compliance perspective—imposed or implied—and improve business conduct. With respect to quality, this has already passed the point of debate, as quality has become an inevitable differentiator for businesses—it is important for businesses to demonstrate how the value offered differs from the competition and relates better to the goods and services being offered. Quality has been defined by ISO 9000 as the degree to which a set of inherent characteristics fulfils a given requirement. The American Society for Quality (ASQ) defines quality as excellence in goods and services, especially to the degree they conform to requirements and satisfy customers. Deming defined quality as a predictable degree of uniformity and dependability, with a quality standard suited to the customer. In all these definitions, consistency and customer satisfaction have been the key messages. Consistency is related to reliability—to be able to produce a consistent result over a period, which is expected to boost customer satisfaction and, in turn, (re)define success criteria for the products and services. However, why quality—objectively or otherwise—has not been defined to valourize environmental considerations is not known, including to reason why reliability or customer satisfaction is not based on environmental considerations. Is it a case of deliberate misinterpretation or reflective of the progressive elaboration of our social conscience?

Seems like a bit of both, as customers have always relied on the products, processes or services to have quality as an inherent attribute, even if that means not incorporating environmental measures that would have improved how these are designed and developed, instead of consciously seeking their buy-ins and fulfil those considerations. For example, a good meal in a quality restaurant would automatically mean quality ingredients, quality processes and quality services, which should not be about openly seeking an a la carte choice from the patrons and offer irresponsibly sourced produce or ingredients otherwise. However, the market economy has rarely been designed to seek inputs from customers on designing products or services or calibrate how they are performing against individual criteria, so quality has been defined as per the convenience of the industries, to the extent to which it could be sold. At the same time, it also points to how quality itself has been defined—it is an output function of the offered goods and services, and *not* an inherent characteristic of the processes, and reflects the inherent nature of collective human wisdom and, in this case, of businesses to act on, react to and shape it. Although green activism has been pushing the envelope for societies to be aware of environmental issues, this is relatively a nascent phenomenon.

Another point relevant to this discussion is the demand for products and services to be environmentally benign, which has given birth to the notion of having products defined as 'organic', 'green' or some other adjective that passes message regarding the choice consumers have to be a part of environmental activism. This differentiation needs some systemic means through which it can be evaluated. Unfortunately, the existing quality frameworks do not offer any specific means of connecting the differentiating factors to the message (read: package) it appropriates (for better gains or value propositions?). This does not stop firms and businesses from searching for a way to advance quality as well as environmental considerations, at least theoretically. But then the onus shifts to the businesses to define a way to implement greener processes and communicate so that the customers can believe their claims are for reasons beyond the usual rhetoric of 'trust me'. It is not unnatural for firms and businesses to strive for that level of excellence, and examples are available as well; but evidently, this striving is yet to become business-as-usual. In contrast, firms and businesses can argue that consumers are not always sensitive or demanding about environmental care, and not necessarily interested in shelling out 'extra' dollars for firms to maintain that level of competitiveness. So, if firms have to institute environmental care as a feature of their products and services, society and consumers should be willing to bear that cost. This questions why institutionalizing a separate framework for environmental considerations is felt as a necessity, where demand for environmental care has not matured into an organic demand. Moreover, if firms do not adopt any institutionalized standard, can they still be environmentally superior and be thriving at the same point of time? Some of these questions are discussed further on.

Any EMS framework aims at improving organizational interaction with the environment in a systematic manner. That needs the business and its processes to be aware of how its processes, procedures, products, people and services are interfacing and interacting with the environment and how they can contribute to enhancing or improving it. Instead of using a rule of thumb, an institutionalized framework helps in establishing the perspective that the constituents of the firm can relate to the goal of environmental well-being, by relying on a normative background. This includes methods, tools and related vocabulary that the firm needs to work with and transform the state of affairs. Scholars have been researching the antecedents and motivational aspects of firms in this area, relying on frameworks from business management, engineering, ethics and other core theories to hypothesize why firms should or would behave in a certain way. These reasons include:

- Fear of losing out to the competition or getting itself indicted due to regulatory or other pressures (*business-as-usual view*),
- Desire to become a champion of environmental concerns, or as a concerned corporate citizen, feeling equally responsible for societal concerns of the degrading environment (*instrumental view*),
- Consciously pursuing environmental strategies due to own dependence on the environment to satisfy business needs, for example, in the extractive and tourism industries (*legitimacy view*),

- Considering an environmental orientation as a business imperative to help position its products, processes and services as being of higher or better quality or capabilities (*light-green view*),
- Working towards improving the sustainability of products and resources because that is the right thing to do and being concerned about dwindling resources that would severely constrain the world of tomorrow (*deep-green view*).

Using empirical validations to examine these needs is one side of the equation; the multiplicity of frameworks existing today is the other area of concern, given adoption of EMS is voluntary. Here is a list of different EMS certification standards (Table 17.1).

However, whether adoption of an EMS would improve the overall environmental impact of a firm is a matter of debate, not to mention whether it enhances the competitiveness of a firm's services or value offerings. This disconnects being environmentally benign from *how* it can be achieved, and points to the certification market, where other than the certification cost, the cost of annual audits, recertification and ongoing maintenance of environmental performance adds to it. The cost of implementing and maintaining one framework versus another would depend on the regulatory needs, if any, and also the provisions of the framework—for example, EMAS is for the EU and necessitates that firms declare environmental aspects and other quantitative data in the public domain, which is tied to transparency. In addition to these complexities, probably three elements remain outstanding and are worth further discussion:

a. Does having EMS frameworks defined by the same body that defined quality systems (QMS) reinforce the basic philosophy of continuous improvement that any QMS propagates, where the implementation of any kind of EMS is expected to drive firms towards continuous and improving environmental care (morphogenetic changes)?
b. Secondly, even when not specified, an EMS brings in a certain level of organizational transparency and helps initiate change in the processes in an ordered manner. Nevertheless, it would remain a systemic burden on the firms, benefits for which might not be quantifiable, and can be an impediment to its adoption.

Table 17.1 *Different EMS certification standards*

Certification	EMS area and details
ISO 14001	EMS: requirements with guidance for use
ISO 14004	EMS: general guidelines on implementation
ISO 14006	EMS: guidelines for incorporating eco-design
ISO 14015	Environmental management: environmental assessment of sites and organizations (EASO)
ISO 14020-5	Environmental labels and declarations
ISO/NP 14030	Green bonds: environmental performance of nominated projects and assets
ISO 14031	Environmental management: environmental performance evaluation; guidelines
ISO 14040-9	Environmental management: LCA
ISO 14046	Environmental management, water footprint: principles, requirements and guidelines
ISO 14050	Environmental management, vocabulary: terms and definitions
ISO/TR 14062	Environmental management: integrating environmental aspects into product design and development
ISO 14063	Environmental management: environmental communication, guidelines and examples
ISO 14064	GHGs: measuring, quantifying and reducing emissions
ISO 19011	Guidelines for auditing management systems: one audit protocol for both 14000- and 9000-series standards together
EMAS	EU Eco-Management and Audit Scheme (EU EMAS) I, II, III

Source: Author.

c. Since demands to become environmentally proactive have not passed through the process of debate by the collective conscience, where discussions and deliberations by different stakeholder groups would have shaped and contributed to the robustness of the theoretical foundations, adoption of an EMS would have to be prescriptive or a part of the regulatory framework; else firms would have to evolve a need to get one.

It is the last point that is examined in the next section, where environmental proactiveness is evolved as a business model.

GREENING FIRMS: EXTENDING THE ENVIRONMENTAL VIEWPOINT TO ORGANIZATIONAL PROCESSES

The literature from the different sources that I surveyed focused on different aspects of EMS, including their implementation, analyses of certification standards, system- and process-related changes, cost of EMS certification using ROI or TCO yardsticks, and outcomes, including competitiveness and market response. However, their positioning with respect to how an EMS connects to the surrounding functions to influence business processes and how environmental information is exchanged across functions and processes remain outstandings on the research agenda. In any case, an EMS is not an operating function of a business; it provides advisory services, depending on the organizational value chain. At the same time, the EMS framework is expected to connect organizational processes, procedures, people, products and services in (re)defining environmental objectives, and to help the organization build a systematic response towards the environment. This needs exchange of information between the constituent processes and the EMS, in order for the EMS to analyse ongoing patterns and activities and help to better realize the organizational objectives. This would result in establishing a common vocabulary and terms throughout the firm and allow data to/from individual functions and processes to capture and encapsulate desired intent. In this book, the area of exploration is how an EMS connects with upstream and downstream functions to capture, codify, record and transmit informational elements that are essential for it to perform its functions while seamlessly connecting to other subsystems.

Case of green supply chain management

Green supply chain management (GSCM) would enable an environmentally conscious supply chain and contribute to the firm's environmental goals of a reduced energy footprint and lowered product miles and carbon emissions, amongst others (Srivastava 2007). Implementation of a green supply chain could lead to the reengineering and design of an environmentally improved GSCM process with feedback loops (with the 5 R's) to improve material and energy intensity of products and services, as exemplified in designing environment-friendly furniture in the manufacturing business (Alvarenga et al. 2013). Accordingly, the physical processes would have to be recalibrated to achieve the desired environmental goals.

Researchers have studied the information needs of GSCM in different periods to reflect on the role of information technology in designing a sustainable supply chain and the ongoing trends of research. Most of the findings from green functions and information system are limited, which point to low or no significance of how information carrying capabilities are designed to connect with green functions. I, however, believe that the role of information flow and information system is crucial in improving the environmental sustainability of supply chain, one that relates to the overall impacts due to movements of goods and services, including the reverse logistics. In an empirical study on pharma companies in Jordan, researchers found the information system to be capable of having a positive impact on the sustainability of a business, where GSCM can be a mediator of information. Compared to that some others considered greening of business functions and information carrying requirements as two separate strategic endeavours. This has been an oft-repeated area of research where the general mindset of firms is to develop information on the environmental aspects and impacts in isolation from the underlying business activities. This is where we leverage a process-centric approach to illustrate the case for GSCM.

To illustrate

Organizational impacts due to different business activities can be analysed as processes that go into defining a business activity or a function. In theory, a business process can be defined as an interrelated set of activities that contributes to generating economic value for a firm (Scheer 2000). So, at the micro level, every business activity would interact with the environment and generate some negative impact. For example:

- Manufacturing of 'A' by processing raw materials that would need 'B' units of energy for conversion
- Dispatch of 'X' tonnes of goods to destination 'Y' that is at a specific distance from the nearest warehouse.

In the first case, energy used to manufacture 'A' could be sourced from captive sources or from external producers. In both the situations, different types of fuel (ranging from fossil fuels to renewable source) might generate negative impacts due to emissions. In the second case, the dispatch of 'X' can be accomplished by using a suitable mode of transport (say, road freight) to generate the corresponding product miles. Accordingly, process decomposition would enable modelling activity–environment interactions better (for example, Figure 17.1 details emissions due to the process X.0.1.1). While the users and decision-makers can go with an unilateral decision to evaluate processes or their environmental embeddedness within an organizational framework within a set of given economic, environmental and legal choices, they would need information on areas that the traditional information system has never been concerned with (for example, emission levels, stakeholders' expectations, geospatial considerations, and others).

This opens up a role for green processes—for example, to handle evolving environment methodologies and information needs, to compute environmental aspects (e.g., IOA, environmental LCA and GHG accounting), and generate information that addresses stakeholders' needs and local regulations better. Accordingly, the process view helps to connect the need for green information with business functions and processes better and brings in the role of users to advance its adoption.

COMPLEMENTARY NATURE OF EMS AND ENVIRONMENTAL ACCOUNTING

Any EMS framework allows the firm to define where and how elements of its value chain and supplementary functions impact the

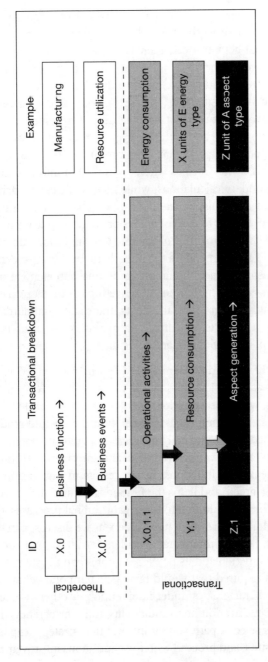

Figure 17.1 *Functional decomposition of a business function*

Source: Author.

environment, and then lets the organizational information system capture and relay these to the desired end. Information captured from organizational activities can then be translated to where and how the aspects were generated, environmental hotspots identified and taken up for remediation, and used to develop an intervention strategy to improve the situation. A case in point is lowering the carbon footprint of the organizational value chain, which would need data and information across the life cycle of products, processes and services, including impacts that these produce at the end-of-life phase as well. This works in cases where EMS-related information is generated via secondary calculations, while the processes are carried out in the business-as-usual manner. The captured information is then relayed to the EMS experts to derive conclusions and provide opinions to help with remediation or improvements. On the input side, this would need any EMS to depend on the underlying business processes from different business functions, an organizational sub-element that is responsible for generating it, a quantitative tool or method that is required to quantify environmental aspects, translation standards to develop equivalent impacts, and finally, a necessary set of units of measurement applicable to the quantitative data.

Consider another scenario, where the EMS participates in redesigning the processes and provides guidance to improve pro-environment decision-making, where organizational processes won't wait for data accumulation and ex-post information generation. This would require a process to embed environmental considerations into its execution systems, including the flow of relevant data and information through different subsystems and functions, which results in organizational processes turning 'green' and helps the EMS in 'greening' the process as well. This is where, in addition to the organizational processes, the organizational information system turns green, as we discussed in the previous chapter. For environmental considerations to be embedded into the business function and its processes, more than a system, systemic thinking is required that would review its temporal behaviour and isomorphic characteristics under intervention or otherwise, and (re)calibrate impacts. In addition to auditability and transparency, this would help firms select a path and an intervention strategy, capture the associated results, and proceed with the continuous improvement that the quality philosophy within EMS should bring in. At the same time, this would help firms in benchmarking their performance against peers and measure environmental effectiveness (greenness) of their processes, products and services.

In both the cases, however, two areas of concern inevitably remain—(a) the temporal aggregation of quantified aspects over multiple process areas and (a) leaving non-production (support) activities outside the purview of EMS and limiting changes to larger or one-time changes. In any EMS, data across timelines, aggregated or otherwise, could prove to be challenging to evolve environmental profile of a process, product or a service. It is easier for any reporting system to report X units of aspects from Y process or processes. However, other than aggregation of temporal data that hides any associated variability from the underlying processes and related inefficiencies and lumps the granularity and could be sending wrong signal. Another aspect to consider is linking the need for information across systems that carry what is needed by the EMS to other sub-systems and vice-versa. For example, for the EMS to be able to weigh in on a capital purchase decision, besides the traditional CBA, it also has to invite the environmental credentials of the suppliers who believe they are qualified enough to fulfil the requirements, as also the environmental profile of the product/service and its evaluation by EMS experts using tools and methodologies that can translate external information to the relevant kind being used within the organization. This brings in the complementarity of EMS and environmental accounting, and the way they mutually engage in organizational improvement.

One unique perspective in EMS literature is the non-inclusion of non-production activities in favour of defining a perimeter around the production, processing or manufacturing environment. The intent here could be to enable in-depth analysis of these processes in the belief that improving these would improve the overall greenness of the firm. However, the facts could not be farther from the reality. For a firm to orient itself towards environmental care, it needs to improve its overall focus and deliberate on concerns holistically. This includes the need for support functions such as human resources (HR), accounting, information technology (IT), MIS, and others also to come together with the core functions and contribute. One of the arguments that is being presented here is that accounting and information system can contribute to this and they do have a role to play, although I would imagine from my personal experience that accounting functions do not usually participate in quality- or EMS-related activities, as accounting traditionally is not seen to be influenced by the activities needed to address quality or environmental challenges—a belief system that is the reason for this text to exist in the first place. Accordingly, other than offering information on expenditures that can be related to these areas (partially or completely), it would make sense to agree that environmental accounting can capture some part of what directly connects to the EMS. This would also lead to cross-fertilization of ideas from two different domains, EMS and environmental accounting, which is nearly non-existent now. Some of this is explored in the last segment of this chapter. Last but not least is the conundrum of one-time change. This is where the light-green version of environmental care differs from the mature, deep-green view. By now we know that the light green view is concerned with short-term benefits or low-hanging fruits that are already available for the firm to cash in on, in place of the more fundamental changes that the organizations should be looking for. Common examples of these changes could include moving energy sourcing from the providers of non-renewable energy sources to those sourcing from renewable energies (solar, wind or hydro). The deep-green vision should question sourcing to co-generation that contributing to the grid is green enough, where the surplus demand of otherwise green grid is met by diesel generator—for example 40 MW regular supply from wind turbines and surplus demand (peak load) of 5 MW is being met by conventional energy supplier. This concerns the detailed information needs of the EMS as well as the information that it needs from other processes to improve, not in isolation, but by connecting to and capturing insights from the underlying processes to define new information. At the same time, the role of human resources in instating environmental care as a part of organizational policy cannot be overemphasized. It supports the involvement of people processes to build the organizational DNA that develops the culture of environmental care as being indivisible from the core objectives of the firm. It also relates to employee motivation and fulfils their ethical need to be a part of doing the right thing and caring for something beyond the self. This is where we turn towards the intrinsic tendency of businesses to turn green—that is, a business model that is environment-centric, and an example to achieve that is explored next.

ENVIRONMENTAL PROACTIVENESS AS A BUSINESS MODEL: CASE OF INTEGRATED WASTE MANAGEMENT (IWM) FRAMEWORK

In this segment, arguments for building environmental proactiveness are examined through a business case that extends beyond the adoption of EMS certification. Accordingly, the role of waste management and its integration within the organization is examined as a business strategy within the hospitality sector, which is not a well-researched area in literature. Research involving IWM practices are generally concentrated around the domains of public policies, solid waste management programmes of municipalities, and isolated studies of waste management practices in the tourism sector. Within industrial organizations, waste management is mostly treated as a backend business function that deserves attention no more than running the errands that the firms produce. In this segment, we theorize that a unified organizational framework that integrates waste management within the operational areas and seamlessly connects them to strategic decisions represent an effective solution: to redesign and contribute to the organizational value chain. This innovative business strategy could be a binding agent (e.g., environmental certifications) or external stimulations (legal and societal considerations) to produce an eco-effective performance (environmental considerations with financial stability), as we shall see.

IWM starts with facility design and extends into the operational aspects of the organization (following a top-down approach). The structural elements of facility design cover building layout and fixtures, while the operational elements cover processes and procedures that are designed to reduce or neutralize the negative aspects of operations. The case study of CHS facilities (covered in Chapter 14) considered structural and operational ingredients that consumed less resources (Table 17.2). A review of the data from Table 17.2 reflects how organizational

Table 17.2 *Eco-friendly organizational measures and their positive impacts*

Eco-friendly strategy details	Avoids	Contributes to
a) Structural: Eco-friendly architectural highlights		
Building plan: north-east facing	direct sunlight	reduced energy use
Façade: extended and protruded	low sunlight	reduced energy use
Roof top: three-layer clay bricks (Coba bricks)	heat absorption	reduced energy use
Timer-based lighting system	power consumption	reduced energy use
Cementation: Portland pozzolana cement (includes 15% fly ash content)	pollution	use of fly ash
b) Interiors: Pro-environmental thinking		
Furniture: MDF	natural wood	saved trees
Plumbing: aerators/flow restrictors	excessive water use	reduced water use
Lighting: LED or PL lamps, T5 fluorescent tubes	high energy use	reduced energy use
ETP: water treatment	wasted water	reduces consumption
c) Operational and guest participation		
Four-bin practice: Garbage segregation at source	mixed-up garbage	100% garbage reuse
Guest amenities: all-natural materials (cardboard hanger, jute bag and slippers)	plastic and synthetic materials	100% reusable
Vermicomposting: recycling bio-wastes	externalities	saleable compost
Eco-nights: Opt to raise room temp. by +2 deg.	increase in AC load	reduced energy use
Eco-tents: Opt not to change room linens daily	linen load	reduced water/energy

Source: Author.

structure (external as well as internal) is connected to the operational framework and processes.

IWM at work

Although Indian organizations (like those anywhere else) are under no legal or statutory obligation to account for their environmental performance, CHS' facilities seized the early movers' advantage to embrace environmental stewardship and developed it as a competitive differentiator within the hospitality industry. A progressive environmental outlook helped CHS to gain environmental superiority as compared to other facilities in the region and provided opportunities to develop a business case for pro-environmental thinking in a country where environmental sensitivity and business interests are yet to converge in a common ideology (Sidhu 2011). To cover all the aspects of waste management efficiently, CHS facilities built an infrastructure that could handle collection, flow and subsequent treatment of the wastes in an integrated manner and assimilated these as part of the individual business processes, detailed hereafter:

Solid-waste cycle

Solid waste was segregated at source using the four-bin approach. The four bins are color coded and marked to enable sorting of waste at source. The green bin was for reusable materials (e.g., paper, recyclable materials and clothes), the white bin for recyclables (e.g., glass and plastic bottles, tins and newspapers), the red one for non-recyclables (e.g., Tetra Paks, butter paper, oily plastics and toothpick sticks) and the black bin for food wastes. Materials in the green bin were reused within the hotel—for example, discarded linens were used for making uniforms, caps, pillow covers and dusters for maintenance activities, and to save expenditure on tissue paper and dusters, the recyclables were sold to recyclers. Non-recyclables were transferred to landfills, while food waste was used for composting. The kitchen steward managed the bins and weighed, stored and recorded waste before transfer and disposal.

Waste-water cycle

The waste water was transferred to the community ETP and the treated water was recycled back into the hotel's flush system. This recycled water was routed separately and used for the purposes of flushing, cleaning and gardening. Practical use of grey-water recycling reduced the load for potable water consumption. To enable that, the plumbing system of the facilities was laid down in such a manner that the recycled water could be stored separately and made to flow through to the designated areas only. The water-inlet pumps were equipped with variable water-level controllers to control the inflow of water based on the level of water in the main tank. This intelligence helped to save energy and optimized the water inventory.

Reduction in energy use

The structural elements of CHS led the facilities to consume less energy (e.g., use of double-glazed glass, LEDs, spots and timer controls). In addition, the heating, venting and air conditioning (HVAC) system was augmented with glycol chiller technology, which used glycol in the chiller to minimize the peak load and reduce the temperature difference between the input and the output fluid. This reduced energy consumption and peak load demand (NSW 2011). Other than that, guests were invited to participate in the eco-night programme, which reduced energy demand on the HVAC system. The heat from the HVAC system was not allowed to escape but was rerouted to heat water, by allowing the water to pass through heated tubes. The hot water maintained a mean temperature of around 60 degrees Celsius, obviating the need for alternate energy sources for this purpose. Also, all exhaust fans (air circulators) and filler or feed pumps were equipped with timers.

Gate-to-cradle cycle of biowastes

The bio-waste collected from the bins was composted using vermiculture within the in-house composting facility, which was developed, maintained and run by the management. The composting process involved cleaning and rinsing the waste, freeing it from acidic contents, and pulverizing it into a thin paste. The paste was mixed with the soil of the pit. Once the process cycle completed, within seven to eight days, the compost was separated and readied for sale. In addition to composting, food waste was also routed to animal farms like the piggeries.

Employee training

The HR department facilitated training to sensitize employees to the importance of caring for the environment while running the facility. Training programmes were organized at regular intervals to ensure employee participation and sharing of ideas. The training programmes encouraged employees to think outside the box as well. Also, the facilities involved employees in social causes such as arranging workshops for schoolchildren and participating in cleanliness drives. These activities generated positive externalities and helped the facilities in the fulfillment of social commitments.

Use of green produce

CHS facilities sourced products that were green. This included green toiletries for guest use, biochemicals for cleaning of the facility, room amenities that are free from plastic or any other synthetic material, using MDF instead of natural wood, glassware and jute rather than plastics and paper.

Ecotel certification as the binding agent

Ecotel is one of the pro-environmental certification standards for hotels and promoted by the US-based environmental consulting firm, HVS International. Certification is provided under the Ecotel branding and awarded after completion of on-site auditing procedures. Recertification and auditing procedures are carried out every two years (Mann and Thadani 2010). The certification criteria covered energy efficiency, waste management and recycling, water conservation, legislative compliance, and employee education (also called the five globes). Certification played the role of a binding agent or stimulant to help the facilities achieve control over the consumption of resources and integrate the waste chain into every process. Collaboration between the facilities and the certification agency created a two-way channel to share expertise and a platform to discuss the environmental impacts of new business practices and operational policies.

Generalizations and further scope

Compared to the manufacturing industry, service industries generate intangible outputs that are delivered to the customers as a bundle of experiences. Waste, on the other hand, is a tangible by-product of service-creating activities. Since CHS operated with an environment-friendly infrastructure and complementary operational processes, the performance of its waste management system cannot be viewed as an isolated outcome of its operational arrangements, a point crucial to implementing the IWM framework. This is different from the causal chain of waste and its traceability that would generally limit the environmental considerations of firms towards end-of-pipe solutions. Based on findings from the CHS case study (section 'Case Details' in Chapter 14) and leveraging a systems perspective, a generalized framework of IWM is proposed (Figure 17.2), one which is capable of instituting morphogenetic or second-order changes in the organizations (Fraser 2012) and help them imbibe environmental care as a primary focus of business, one that is not subservient to its economic performance.

Accordingly, the performance of CHS can be viewed through the IWM framework to present these generalizations:

- It requires upfront thinking by the management to steer the organization towards lowering environmental impacts and invest in the infrastructure necessary to institute and develop practices that would reduce the consumption of resources and the generation of environmental aspects. Evidently, internalization of environmental care transfers the cost of infrastructure development from

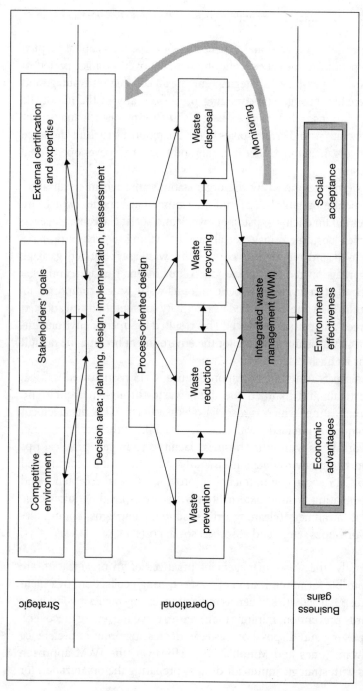

Figure 17.2 Integrated waste management system

Source: Author.

operations (operational expenditure, or opex) to capital (capital expenditure, or capex), offering benefits over a longer period of time. Capex-based strategies generate an economic advantage and benefits through standardizing processes early in the life of the operations, whereas opex-based short-term waste management schemes would create spikes in the operational structure and force the organization to alter its tariffs and prices to recover operational costs.

- Externalities saved by the organizations through internalization generate tangible social and environmental benefits, which are helpful for eliciting customer loyalty and societal acceptance, even when customers may not always be willing to pay extra. CHS reused and recycled its entire solid waste save for red garbage (garbage in red bin where red colour means garbage cannot be recycled), recycled 100 per cent of its waste water, composted its entire biowaste, minimized its electricity consumption and adopted green and natural products. This resulted in superior environmental stewardship, and care for the environment become the default for its business.
- As a part of its social responsibilities, CHS participated in local festivals and handled the floral waste through recycling and composting, which is a social overhead. The CHS management frequently interacted with schoolchildren from the locality and ran workshops with other business facilities to impart practical tips on being environmentally friendly. These activities contributed to CHS' social commitments, improving its social acceptance and generating positive externalities by way of spreading knowledge and awareness (cleaning drives, school campaigns, community workshops, etc.), and absorbed social costs.

Generally, the waste management practices of an organization are responsible for minimization of wastes through collection, treatment and disposal activities, whereas mitigation arrangements involve costs towards prevention, minimization, reuse, recycling, energy recovery, composting and disposal of waste, in decreasing order of preference (Radwan, Jones and Minoli 2010). However, the IWM approach starts with strategic inputs for design, preparing the organization for improved utilization of resources and prevention of waste, not to mention building conscientious organizational processes. These efforts better embed environmental thinking within employees, processes and services, and contribute towards developing it as a competitive differentiator for the business, thereby converting waste management into an inherent part of organizational success. In a business-as-usual scenario, such embeddedness would obviously be lower. Learnings from CHS include the feasibility of integrating environmental considerations into the organizational framework and could be strategized to seize the early movers' advantage in different areas. Even though such forward-looking business ideas are yet to become a commonplace, the study established the competitive edge that it offers to the interested firms.

Environmental stewardship by an organization holds it liable towards superior environmental performance. However, instead of using CBA to calculate incremental dollar investment for measuring incremental business gains, the management can use an IWM framework to steer organizations towards a value-driven approach. Integrating environmental considerations as the primary focus of business includes adopting environmental embeddedness as an integral strategy of business and develops care for the environment as a structural component of its competitive differentiator. Advancing this argument further, IWM helps businesses to adopt waste management as a part of their strategic thinking and reposition every process to minimize environmental and social externalities, thereby helping the firm to demonstrate the highest level of environmental stewardship. On the negative side, even though IWM generates tangible savings in the form of better utilization of resources and reduced levels of waste, the cost implications of environmental certifications and infrastructural investments are impediments to adopting such a framework. Still, firms can develop environmentally conscious behaviour due to policy initiatives like tax rebates and interest credits. There is a window of opportunity here for policy makers to develop policy instruments like differential tax treatments, tax holidays and other mechanisms that could act as catalysts for firms to adopt higher environmental standards.

Sustainability and Environmental Interfaces
Recent Advances

INTRODUCTION

This is the final chapter of this book, concluding the journey that we have taken. For as long as nature and everything within it has been believed to be a self-contained system, philosophically and scientifically, the sphere of human interactions over different eras has been studied as sub-phenomena confined within the development themes of that age. At no point of time have we ever found these events to be epoch-changing or difficult for Nature to deal with, other than the ones resulting from human endeavours after the Industrial Revolution. Moreover, scientific developments and machine culture proved counterproductive to maintaining a symbiotic relationship with nature. This should lead us to ask whether expanding the industrial ecosystem is a risk worth pursuing. We have seen that although contemporary advancements and ecological challenges are contributing to the expansion of human knowledge, we are still miles from evolving into ecologically sensitive communities. Our capture of nature–human interactions is hardly significant as new facts and perspectives are emerging, thanks to advances in ecological sciences, particularly the ones having genesis in collective human efforts. We connected these and other concerns to explore how economic activities shape industries, now and in the future, and considered whether they are transdisciplinary in nature. Drawing on economics, business management and accounting sciences, contemporary developments were examined in firms contributing to sustainability. Accordingly, a comprehensive review of the accounting sciences from the perspectives of accountability and environmental concerns discussed how accounting has traditionally remained confined within the economic affairs of firms and why a new framework of accounting would be meaningful. We have now reached the point where new advances in related subjects can be reviewed. These are the boundary objects that can and will impact the discipline in times to come. Moreover, our belief that the accounting sciences are certainly capable of further contributions is fundamental to exploring these advances and their relatability to the sustainability sciences.

THE CHANGING ROLE OF FIRMS IN THE LIFE OF *HOMO ECONOMICUS*

Once we accept that business firms are the progeny of our collective economic aspirations, we have to acknowledge that the economic system and our interactions with it have been impacting the environment, our social structures, and our lives and thinking. Here is an example, to start with. I read somewhere that after the invention of colour television, more people started dreaming in colour (Schwitzgebel, 2002), which points out human susceptibility to the technological advancements and changes happening all around us. Corresponding to this is how our lives are influenced and shaped by the business world. For example, firms use psychological profiling to target customers or use nudging to push them towards products and services that customers believe they need. Organizations originally purposed to serve the needs and wants of society are now insidiously shaping our desires, habits and needs, not to mention our lives, in a way that separates us from the ground realities of life, absolving us of the need or want of knowing how goods and services are procured, produced and manufactured—like the neatly packaged meat products on the shelves of supermarkets isolating us (thankfully!) from the miserable feeling of knowing that the animal was alive a while back, drying up our compassion and gratitude, replacing it with purchasing power, our ultimate saviour!

While the economic system is an abstraction of how trade and commerce involving goods and services are transacted, its sticky character does not remain confined to numbers and trends but transcends to include us and everything as well. This stickiness was not so uncomfortable until the economic machinery and its processes got into our thoughts and decisions to the extent that we started using economic rationale to define and live our lives. This is when economic thinking became a part of our collective conscience. This is not only true for an individual but also reflected in our collective choices or behaviour. For example:

- Examining the suitability of a job solely based on its remuneration reflects the economic rationale of the decision-making process at an individual level.
- Selecting eucalyptus trees for afforestation programmes on the Himalayan slopes or allowing extractive industries to stake mining claims on the bauxite-rich hills in Niyamgiri in India (Sidhu, 2011) without investigating the long-term impacts on aboriginals who have been living there for centuries and whose faith regards the hills as living gods characterizes economic thinking at social and policy level.
- Early colonization of distant countries by European traders and the destruction of aboriginal life and social systems in these occupied territories to create new markets were not the outcomes of the Industrial Revolution but resulted from economic thinking that suppressed every other rationale argument that did not fit the paradigm of economic merit. Today, these colonizers have been replaced by international economic bodies (the International Monetary Fund, or IMF, and the World Bank) who would turn every occasion to support a Third-World country into a market opportunity for the global north.
- The cost of education is now synonymous with some arbitrary ranking of educational institutes, which directly corresponds to the brand values that markets and organizations care for, while students are not assessed by their future potential, but as a current percentage of the GDP, based on the students loan (Li 2013, and other similar writings).
- Money or currency, which was initially a medium invented by societies to measure the value of certain goods or services in a

purely arbitrary manner, has come a long way from being required for enabling a commercial transaction to occupying a prominent, even defining, position in our lives. The concern here is that the lives of the human species are being altered not due to economics alone but by instituting economic measurements in everything that we think about, believe in and act on. This institutionalized system has become a central theme of our lives.

- The Olympics is no longer about celebrating the hendiatris '*Citius, Altius, Fortius*', but more of a sponsorship circus and competition for media rights, where economics keeps on beating human merit. Moosa and Smith (2004) found GDPs, population, national expenditure on the health system and number of athletes are good explanatory variables to predict medals tally, out of which three translate to economic equivalents.

The point relevant to decipher here is how corporations and business practices have changed over the years and emerged as indivisible from our lives, impacting societies and human thinking over time and creating a rich yet debatable anthropological narrative of our social institutions, While businesses, money and the market have played a unique role in economic development, these have expanded their influence to human intellect, leading the overwhelming majority to view certain things in a certain way. This has instituted a system that misses the ground realities, measuring things through money, ultimately transforming us into *Homo economicus*, disconnected from nature and reality. What can be more challenging than that?

ECOLOGICAL SCIENCES AND SUSTAINABILITY

Although ecology or 'the study of the house' was proposed in the late 19th century by German scientist Karl Möbius as biocoenosis and later as geobiocoenosis by a Russian scientists, it was not until the 20th century that it formed an independent discipline, coming out of the shadows of biology or the environmental sciences. Ecology has an overlap with all these subjects and more and, in that sense, is transdisciplinary in nature, as against the interdisciplinary orientation we may believe it to be, more so because it draws upon the fundamentals of multiple branches and studies life processes, their adaptations and interactions with the biotic and the abiotic worlds to develop a basic structure that can explain the evolution of life and the earth as a living planet. This includes the study of fundamental units of all life forms, their evolution and sustenance in different trophic or energy levels, their resilience and mutual dependence on the non-living world to sustain themselves and evolve (Odum and Barrett 2005).

The ecological sciences have relied on the laws of thermodynamics to explain energy and its transference across different life forms and abstracted the principals of general systems theory to explain diversity and ecosystem behaviour. But why is the study of ecology relevant in this context? To me, the ecological sciences break away from the traditional mould of science that has depended on the deductive approach (using one unit to explain the behaviour of the entire system) to use synthesis (the sum of the parts is lesser than the whole). Another area of significance is its search for a fundamental way to explain the symbiosis of the biotic and the abiotic worlds, where any one side does not outweigh another. This is somewhat closer to the ecocentric approach that sees every natural system of equal value. Also, the deductive approach of conventional science has been tested and found wanting when applied to life forms, large-scale problems and behaviours of the biotic and abiotic worlds to sustain life. The ecological sciences have been developing diverse frameworks to integrate the mechanistic and evolutionary properties of biogeosystems and ecosystems in a bid to establish general principles that explain how the entire spectrum of life systems dances around energy and matter, constantly changing forms sustain one another and co-evolve.

More than anything else, this has led ecologists and breakaway thinkers to move away from the expert-led approach of conventional disciplines and to synthesize learning from different disciplines. One of the breakaways in this area is to propose the common unit of 'energy' as being fundamental to transitioning from lower life forms to the upper ones. Using energy as the fundamental currency of natural interactions, ecologists explain how natural as well as anthropocentric systems operate. For example, the sun's abundant energy is consumed by plants through photosynthesis to generate food for the herbivores and oxygen for all living systems, while also consuming harmful carbon dioxide in the process. On the other hand, herbivores get consumed by the carnivores, and/or by tertiary or advanced life forms. The same cycle is followed in water and aqua systems as phytoplankton consume and transform energy to sustain the marine biosystems. In the human world, energy as currency can also explain the built environments of cities and villages, where primary producers use energy to generate a stock of food (energy embedded) that gets transported (energy consumed) to cities. Cities consume food, materials and energy that are brought over large distances by paying for them through artificial energy sources, such as fuel, as they have limited ability to absorb the primary energy sources of nature (such as sunlight) to sustain themselves. Accordingly, their dependency on the secondary sources of energy will keep on increasing till they hit the diseconomies of scale. Diseconomies of scale (which economists do not like to talk about) comes into the picture when the increase in the size of a settlement pushes for more than a corresponding incremental cost of energy, forcing the settlement to develop alternative sources of energy, which would hopefully be to tap into solar energy. This groundbreaking contribution is owed to H. T. Odum and related works spread over the second half of the last century (Brown and Ulgiati 2004).

From the ecological perspective, how do we define sustainability? The ecological sciences define sustainability as the carrying capacity of the earth or any ecosystem at a specific level of complexity. This includes the ability of the ecosystem to be able to maintain different life forms and their interactions within the biotic and abiotic worlds, maintaining overall equilibrium or balance. However, the health factor of a specific species is defined as the space needed for the natural growth of a species under the ambient conditions (refer to the Michigan deer experiment). This would include maintaining the flow of resources through the system in a way that sustains everything within the system. From the ecological perspective, sustainability is not just about fulfilling human needs, but a state of the entire ecosystem and its constituents to sustain life activities. However, how the sustenance level shifts at different population or trophic levels is still an open question. In that sense, sustainability is not a point on a scale but a range of values, within which the system tries to maintain the balance between local highs and lows. How that connects to the global high and lows is also subject to investigation, to the extent I understand.

Extending this to economics reveals why continuing economic growth cannot be a possibility, why this is in turn relevant to our economic policies, and why it is important for our cities to efficiently and effectively process and recycle waste. At the same time, it becomes easier to understand that the market economy shapes our common interaction with nature through industries and businesses to extract (read: extort) benefits from natural systems, but without due consideration to the different services that the ecosystems already provide. The application of a monetary system comes into the picture only when the produce is ready for the market, while all other services provided by the ecosystem are discounted. To rephrase Col. Nathan R. Jessop from the movie *A Few Good Men* (Reiner, 1992), this is what Nature is saying to humankind:

> *I have neither the time nor the inclination to explain myself to you humans who rise and sleep under the blanket of the very comfort that I provide, then brag as if it was you who provide support for my existence. I'd prefer you just said thank you and go on your way. Otherwise, I suggest you sustain earth and life forms on your own. Either way, I don't give a damn who you think you are or what you are capable of....*

Surprisingly, while nomads and aboriginals have been able to get it right through the centuries without sophisticated knowledge, societies and economies that claim to be the developed ones and leading the knowledge world do not get it fully and are continuing on the path where we have ended up today. The global south, in contrast, is fast learning the ways of the global north while unlearning the basic concepts of life sustenance that were ingrained in them not so long ago. This is a double jeopardy, as they say in legal terms, for Nature[1]

---

[1] Double jeopardy in legal terms can be best explained as being placed at jeopardy twice. In the matter of ecological challenges, the global south is not only moving

to deal with. Another element of the ecological sciences is diversity or biodiversity which supports multiculturism, letting all life forms grow and live that adapt to the abiotic world that co-evolves with the biotic one, based on the geophysical nature of the surroundings, as against humans, who go to excessive lengths to promote monoculturism. The human inability to accept diversity goes against how Nature works and stops us from adopting good things that others have to offer. Intrinsic to sustainability is also how the abiotic world participates in the biotic world through different cycles of water and other natural elements (carbon, nitrogen and phosphorus cycles), and through the participation of plants and microorganisms to convert natural elements into the nutrients needed for life systems and to again recycle waste into the natural elements, enabling energy flow through the trophic levels (Odum and Barrett 2005).

At a philosophical level, this is a cosmic dance in which matter and energy participate together to form, deform, reform, evolve, devolve into and revolve around what we call 'life', sustained through both the biotic and the abiotic worlds. This is reminiscent of the cosmic dance of Hindu mythology where Shiva (energy) and Parvati (form and matter) have been playing to dissolve their individual selves to अर्धनारीश्वर in Sanskrit (*ardhanārīśwara*) and create nature as we see it on earth and expand the cosmos. The concept of Ardhanārīśwara, literally to meaning half female (nature) and half god (energy) that coexist, originated in Kushan and Greek cultures simultaneously and the iconography evolved in the Kushan period (30–375 CE) (Flinders, 2007). Instead of discussing religious details, which are not the subject matter of this text, I am more inclined to put forth the philosophical perspective where the simultaneous existence of energy and matter coming together to sustain, dissolve, shape, and exist in creation. Although classical Western thinking would not find this a rational paradigm, it has existed in fables for humanity to embrace for a long time. As to my readers, this is not about promoting any religious doctrine, but rather an allusion to the wisdom in ancient

away from sustainable ways (first jeopardy), it is also fast adopting a materialistic culture (second jeopardy).

practices that can be interpreted as narrating ecological sustenance in a profound way, and I am sure extant in other forms in different cultures as well.

ECOLOGICAL ECONOMICS AND ECOLOGICAL ACCOUNTING

The framework of economic policies and choices has unfortunately promoted a mechanical view of the world, the rules of which are segmented or specific, as we learnt through this discourse and which leave the complexities for heuristics to work out. This seemingly simple world of demand and supply would be challenged if economics were more ingrained in real life and the environment. This is where the disciplines of ecology and economics come together to form ecological economics. While ecological study of the Anthropocene and its interaction with other spheres is a new field, advances in this field also include the study of human–nature interactions at aggregate levels. The latter is the primary focus of ecological economics, encompassing more than the traditional resources of economics or environmental economics, as the latter is more concerned with environmental degradation due to spillage effects produced by economic agents (van den Bergh 2001). This section discusses the developments in ecological economics and how it is embracing diverse viewpoints on ecological sustenance and the impacts of the Anthropocene. However, the threads of development are tangled and the taxonomy is yet to firm in this area.

Although constancy of natural capital has been the bedrock of the ecological perspective of sustainability, its interpretation in places has crossed the economic view to create the ecosystems services view. The constancy of natural capital referred to is related to maintaining natural assets from activities that leverage these so as to prevent altering sustenance permanently. From an ecological perspective, ecological wealth connects these two perspectives. One being constancy of natural capital and another being ecological services they offer (Barbier, 2013):

- The view that is based around *natural capital* holds that the ecological abundance of untapped capital—from the sustainability perspective—needs to be maintained. Although capital relates to the economic view that a firm borrows to invest in assets and create production capacity, while the provider of the capital is rewarded with dividends for the risks undertaken to finance the venture, yet capital is also a source for augmenting the firm's capacity for its future sustenance. Capacity that can be leveraged for production activities and is to be used repetitively or cyclically to generate profits or a flow of returns will need a periodic charge (depreciation) to create a reserve (depreciation reserve) which can replenish the original investment and improve the capacity created in the first place. For economic sustenance of the business, the capital needs to remain constant or improve in due course of time, and pay for inflation as well.
- From the ecosystem perspective, nature and natural systems can be categorized according to the services they provide, which can then be tracked for the impacts that human and natural activities can have in the long run. According to this view, ecosystem services are categorized as four different services—provisioning services, regulating services, cultural services and supporting services. However, from weak sustainability perspective, viewing ecosystem merely as a set of services is closer to anthropocentricity, where nature and natural systems can be viewed as a source of certain capacities or services for human existence. In contrast, ecocentricity would view ecosystem services as a set of services that nature and natural systems perform for sustenance of the natural world.

Advances within ecological economics relate to the impacts of economic activities on the ecology and evaluate changes to, depletion of or loss of ecological facilities or services; they also seek alternatives to reduce the impacts. Research in this field is progressing to evaluate, classify and possibly monetize ecosystem services that impact human activities, and learn from the findings. For example, a lake is not only considered a natural asset (in ecological terms), it is considered as a set of ecosystem services that it offers to humans and others, and if the lake is required to pave way for human settlement, the kind of loss that it would lead to has to be taken into account, including the losses of a water body, aqua services and all other sustenance that the lake provides. By bringing in some form of quantitative evaluation that the economists can use to influence policy makers, this discipline tries to improve the situation, as against letting natural resources and ecological infrastructure be converted into opportunity costs. This is a limitation of the economic view and a market-based understanding that ecological economics aims to improve upon, including developing appreciation towards the loss of natural resources due to human endeavours.

In comparison, ecological accounting is a conceptual framework to apply the methods and techniques of accounting to firm–ecology interactions. Drawing from the advances in ecological economics and systems ecology as well as the quantitative framework of ecological economics, it is expected to be more encompassing than environmental accounting and to capture how firms—as Anthropocene entities—interact with the ecology or with ecological systems. Ecological accounting is also about ecological resources and their utilization, and includes spatio-temporal changes in the resources' profiles. This is an improvement from the view of input–output flow that traditionally characterizes an economy. Accordingly, ecological accounting at a macro level would be concerned with improving the state of green national accounting and to widely cover all ecological resources and the services (the quantitative view, per Ogilvy 2015). Compared to the macro level, ecological accounting at the micro or organizational level would be more related to accounting impacts that a firm produces due to firm–ecology interactions and include impacts on biodiversity, flora and fauna, and local or proximate ecological reserves and assets, and environmental interactions (second- and higher-order impacts) that are beyond the coverage of environment accounting. For example, if a construction firm starts a large-scale housing project within a proximate lake zone, ecological accounting should abstract the ecological concerns this might lead to—or, for that matter, how extractive industries and megastructures like dams and hydropower projects affect an entire region—and to handle the ecological disturbances that these would generate during the entire lifetime of the asset. One such effort can be found by Sullivan and Hannis (2017).

Due to a limited view of ecological disturbances, societies have had a narrow focus on the impacts of megastructures and technocratic solutions that result from policies where short-term gains get the upper hand. Even when the issues are debated via environmental impact assessments or other regulatory mechanisms that exist for competing interests or to enter a polyvocal debate, the economic benefits of today or in the short-term are found to take precedence in the policy circles. Ecological accounting is expected to handle these concerns or at least seems promising for improving conditions by helping firms meet commitments in the wider interest of society. How? Unfortunately, accounting research is still in its infancy and as yet unable to offer any deeper insights into how this can be achieved. Definitely it has to move ahead from the myopic world of the IFRS, which has been totally inadequate to the ecological challenges of tomorrow (Richard 2017). In my extensive survey of the literature, I came across only a few articles that had moved past environmental accounting. Another aspect to work on would be ascertaining the cross-domain knowledge of accounting, economics and ecological sciences that the research is yet to straddle and which would need experts from multiple domains.

How would ecological accounting promote sustainability? Even without being able to imagine how ecological accounting would finally shape, promoting sustainability excluding accounting sciences or any form of it would be equivalent to navigating open sea without any navigational device. More than questioning how firms would enact some version of it to help society believe that they are promoting sustainability, society must acknowledge that they are equally responsible for shaping the ecological responsiveness of firms and industries and in developing policy choices to match the rate of ecological change. What ecological science has been able to prove beyond a doubt is that the sustenance of the ecology is complex and that the web of life is highly interconnected and responsive to everything that goes on around it. At the same time, life-supporting mechanisms at different trophic levels are different and instead of having a definite value of underlying variables to represent sustainability, it is more like a vibration (a range of values) that attempts to kind of resonate with the changes, countering adversaries. If changes persist to such an extent that the threshold is broken, that area or regional system tries to reach a new equilibrium, irrespective of whether that new state might support or prove poisonous to life systems. If things are left to themselves, changes in the habitat can rejuvenate it to support a higher level of diversity and ecological activity, but that is a topic for the ecological sciences to deal with. The accounting perspective has more to do with the accountability of our decisions, beyond the human needs that an organization is a part of, and our learnings from nature on how to keep it alive—for example, by developing a circular flows of goods and services, by developing multiple levels of trophic (or firms) that can operate and sustain itself (from subsistence onwards to transnational organizations) through interconnectivity and interdependence that is natural, instead of abstracting everything to a generic level and representing it as function of a select few variables, assuming a linear relationship amongst these.

SPACE DEBRIS: EXTENDING ACCOUNTABILITY BEYOND THE EARTH

This subsection relates to the human responsibility for and accountability to the celestial world, beyond the earth and nature, and including space, the moon and Mars, which humanity is reaching out to, sending out and planning space expeditions of different types (manned and unmanned). These expeditions and our responsibility for them are referred to here to take cognizance of human acts towards the cosmos in the grand scheme of things, and at least towards the proximate space surrounding the earth to begin with, the latter being relatively accessible. In this regard, recent developments have increased human capabilities to venture into outer space, and in the years to come, they would advance further, making celestial bodies and space susceptible to human activities and the resulting debris that we are or would be creating. Space debris refers to the orphaned materials left (uncontrolled) in the immediate space surrounding the earth and are essentially man-made. International space treatises and the involved agencies are yet to define it officially and bring about acceptance of their legitimacy amongst spacefaring nations. More than anything else, space debris represents our lack of control on the rear end of the technology developments that generally relates to tunnel vision of engineering marvels, without ownership and diligence towards unwanted outcomes. We carry the undeniable legacy of using everything that we can or would like to, altering things without looking for a natural order or learning from what already exists to achieve certain goals or aspirations that are born out of our curiosity, and leave everything else in a state of higher entropy, believing it to be beyond our responsibility. We are euphoric about our common future, without fully realizing our shared responsibility towards it.

Space debris can be created due to objects getting discarded during or after space missions (e.g., inactive satellites, rocket stages, etc.), microparticulate matter (paint and other metal fragments) or generated by a collision between existing pieces of debris (fragmented debris). Objects in space move at hypervelocity and that increases the damage from the impact of collision, even with smaller objects, significantly. Collision between space objects may lead to further fragmentation of these and increase the debris population and, in turn, become a risk to the active space missions. The current debris population is estimated at 22,000 tracked objects that are larger than 10 cm (Aeronautics and Space Engineering Board 2011; Liou 2011) and 150 million other untracked objects (Lehnert 2011). However, the debris estimate may vary with time, source and orbital band (distance from the upper level of the atmosphere) and this is a threat to the peaceful use of outer space. Study of the creation or mitigation process for debris—that is, finding technological solutions—is not the agenda of this discussion. Instead, the idea is to urge cognizance of the pollution of the space environment that we are creating and the lack of accountability of spacefaring nations and their agencies towards this growing threat.

Though the USA is voluntarily monitoring debris (bigger than 10 cm in size) using the US Space Surveillance Network (SSN), the threat of space debris has forced the spacefaring nations to take measures to secure current and future space missions (National Reserach Council 2011). Tracking, managing and controlling the debris is a complex affair, mainly due to technological concerns and the inability of the prevailing space treaties to achieve international consensus and convergence on the risks of a growing population of space junk (Williams, 2008). Natural degeneration of space debris (due to the earth's atmospheric drag) is a time-consuming process and fragmentation above the earth's atmosphere adds to already existing debris of lower particle sizes, posing significant risks to space programmes. The time it takes to completely degenerate debris depends on its initial shape and size and its orbital distance from the earth. Scientific studies show that the orbital debris can remain in space for very long periods, from years to centuries and beyond (Lehnert, 2011).

Though space programmes for now are mostly state-controlled, that does not rule out their commercial exploitation in the future. We are nearing that time, as private companies have already started envisioning short- and long-term space ventures, and working towards their commercial viability. SpaceX from Tesla and the Sierra Nevada Corporation are examples of these, and more will follow suit. Even today, US space agencies involve the private sector to gain the advantages of efficiency and scale of operations (US Congress, Office of Technology Assessment 1990; Jacobs 2011). If we consider a futuristic business scenario of commercialization of space programmes and the participation of private space carriers, how would the debris generated be accounted for under private partnerships? Keeping financials aside, any damage to the space assets and space missions and any liabilities arising out of collision due to debris would call for a different kind of thinking to cover the risks.

The existing literature on the commercial aspects of space debris is yet to delve beneath the surface of legal complications that require international cooperation to develop acceptable platforms for accountability. However, exploitation of space environment and population growth of space debris are not well-defined functions of improvement in legal recourse or time. This means we cannot predict with any degree of accuracy that the growth in space debris corresponds to elapsed time or counterbalanced by the advancements in legal structure. Even without the hypothetical business propositions, potential business opportunities exist to invite private parties into the clean-up process for space debris (ibid.). From the perspectives of our core interest, the requirement would be to establish accountability of countries, space agencies, contractors and other involved entities for space debris. Considering only a small percentage of space debris is catalogued, the absence of international cooperation or central legislation makes this transboundary issue a complex one, well beyond the realms of prevailing practices.

Still, this leaves the question of how accounting is related to it. The multidimensionality of accounting that is proposed here could be a valuable framework to examine space, which the author believes would be an effective arena to represent the collective interests of a multitude of stakeholders and to ascertain the responsibilities of multiple agencies collaborating in a complex endeavour. This, theoretically, can liberate accounting from unidimensionality to evolve into a multidimensional construct. This, in turn, leads to the possibility of innovative thinking to develop solutions and fix responsibility. However, all these would mandatorily require international cooperation and acceptance, and a respectful attitude towards owning responsibility for clearing up the mess that is inherent to our way of life and how we interact with nature—in other words, our shared accountability!

APPENDIX
Mathematical Modelling of Complex Waste

Industrial growth and the human population boom are pushing up the global demand for energy. With less than optimal use of regenerative resources to support this demand, our dependency on coal as the basic fuel source is likely to grow exponentially. This would also contribute to the growth of coal combustion waste (CCW). However, the environmental impacts of CCW—commonly referred to as fly ash (FA)—and its disposal practices have camouflaged the hidden costs and social and environmental externalities into insignificance. Moreover, studies covering the externalities of coal-based thermal power plants have been sensitive about the externalities related to coal mining, transportation and consumption but with less inclusiveness towards the externalities of FA. The same is the case in the extant literature where emphasis on mitigating risks associated with FA is miniscule as compared to the life cycle of coal (for example, Epstein et al. 2012; National Research Council 2010; Xiaoye et al. 2013). FA is also not classified as hazardous waste, though its revised status of 'special waste' accorded in 2010 by the US Environment Protection Agency (USEPA) has brought attention to the scale of the impacts it produces (EPA 2010), but this is insignificant and mostly muted, a *fait accompli* in the life cycle of coal.

FA AND ITS BEHAVIOUR

Fly ash (FA)—a commonly occurring residue during the burning of coal—is a harmful and chemically complex, quantitatively significant by-product produced by coal-based thermal power plants and is one of the coal combustion residuals (CCRs). Here are some statistics: Presently, 70–75 per cent of the installed capacity of India (90,000 MWe) for electricity generation is from coal-based thermal power plants, same as in the USA (National Research Council 2010), which produces 80–100 million tonnes of FA per year. The coal in India is generally of poor grade and produces 40 per cent of fly ash on burning (Asokan et al. 2005; Shamsad et al. 2012).

The uniqueness of FA lies in its ability to fragment into its constituent elements upon disposal, which ultimately increases the concentration of different metal and non-metal compounds in the benign environment. Main components of FA are oxides of silicon, aluminium, iron and calcium, with lesser amounts of magnesium, sulphur, sodium and potassium elements. Other metals and metal-like elements are also found in trace quantities—arsenic, cadmium, beryllium, tantalum, nickel, manganese, chromium, selenium, zinc and other metals (Rowe, Hopkins and Congdon 2002). While the literature views FA as a serious environmental concern, the common disposal practice of accumulating it into ash banks and ponds in dry or slurry form has not evolved significantly (Asokan, Saxena and Asolekar 2005; National Research Council 2010). The ash banks and surface impoundments may use a clay liner, polyliner or even no liner in some cases, as the disposal rules for developing ash banks have not been consistent across regions and countries. This brings us to the contamination and environmental impacts that FA produces upon disposal.

FA and soil contamination

The acreage of land used for an ash pond or dump would remain a wasteland for a considerable period of time due to the excess of metal contaminants. Although the lined ash pond is relatively more prevalent in the US, lack of a similar practice in developing countries such as India can be attributed to higher acceptance of FA's supposedly benign nature (Sushil and Batra 2006). In either case, research is yet to explore the regeneration of land covered under ash banks, which is practically of no use for a long period of time. The available literature is divided in its opinion on the soil contamination they result in. Scientific experiments have substantiated that the richness of minerals and oxides in FA could potentially improve the mineral composition of soil and that controlled treatment of soil with specific percentages of FA have indicated an increase in soil fertility and yield of crops. The results have been consistent in the range of 20 to 30 per cent of FA being mixed in (Arivazhagan et al. 2011; Pandey and Singh 2010; Sharma and Kalra 2006). However, the chances of metal elements entering the food chain due to their increased concentration beyond the safe limit would need further investigation (Singh and Pandey 2013).

Water contamination and impacts on aquatic organisms

The contamination of flowing water due to FA slurry or sludge has not been adequately covered in literature, other than their leachating and accidental release into aquatic bodies. Studies covering the contamination of water bodies through its leachating into the ground have covered multiple pathways such as surface run-off and underground leachating (Singh et al. 2007). Studies have also shown that the proximity of the water bodies to the ash pond is a factor in possible contamination of water and aquatic life (Prasad and Mondal 2008; Rowe, Hopkins and Congdon 2002; Ruhl et al. 2012). As compared to large-volume water bodies, small lakes and water bodies show earlier and larger degrees of contamination from CCR effluents, but site-specific studies are needed to cover the specificity of the damages (Ruhl et al. 2012).

Groundwater contamination

Groundwater contamination is one of the environmental damages that can be traced to the leachating of the metal elements of FA. The extent of groundwater contamination would depend on variables like distance to the receptor well, average depth of the groundwater table, percentage of the population living near the contaminated well, site-specific variables such as the type of ash (conventional CCW, CCW disposed of along with coal refuse, FBC waste), type of liner used (no liner, clay liner, composite liner), soil texture, aquifer type, groundwater temperature, climate, hydrological properties of the region, surface-water type, flow conditions, etc. (EPA 2010). Contaminated groundwater could result in the loss of water resources for humans, while ingestion of water from contaminated sources could result in an increase in health issues, which would depend on the type of contamination (e.g., arsenic, lead or mercury contamination) as it might not always be linked directly to the source.

Health issues for humans

The health issues due to FA exposure for the human population could originate from different routes such as drinking contaminated groundwater or surface water, ingestion of contaminated food items exposed to such contaminants, and direct contact with the contamination on the surface. However, such risks would also depend on the size of the population in immediate vicinity of the ash pond, reference doses (RfDs) of different elements, and carcinogenic compounds that humans are constantly exposed to from different media, such as arsenic (V) and lead sedimentation or selenium (IV) surface run-off (EPA 2010).

These impacts indicate a complex web of ecological interconnections and the need to understand the functioning of the ecosystem better.

ECONOMIC AND ENVIRONMENTAL POLICY EVALUATION TECHNIQUES AT THE SECTORAL AND REGIONAL LEVELS

Direct and indirect valuation methods have been a part of environmental economics to assess the externalities of waste and could be used to improve objectivity in decision-making while dealing with social policies. For directly identifiable impacts of waste, the hedonic price method, travel cost method or contingent valuation method is generally used, whereas indirect methods include using replacement costs or the preventive and human capital approaches. While direct methods use cost as an indicator of fully functional but unavailable ecosystem benefits, indirect methods are dependent on the prices of environmental gains and people's willingness to pay for them and enjoy the benefits. These methods have been used to assess losses at the sectoral and regional levels—for example, estimating losses due to the Prestige oil spill (Garza et al. 2009), healthcare costs due to air pollution in a number of studies (Dorbian, Wolfe and Waitz 2011; Rabl and Spadaro 2000), and environmental impacts due to climate change (Anthoff and Tol 2010; Ortiz et al. 2011). The impacts are established using market-oriented views and by considering the ecosystem services, which cautions the importance of considering ecosystem losses not just scientifically, but also in ecological terms, and move away from business-as-usual approach (Atkinson, Bateman and Mourato 2012; Schultz et al. 2012; Spash 2008).

Advances in research on ecosystem valuation have been based on the scientific unravelling of ecosystem services that capture partial or complete loss of such services. While the underlying theories rely on ecosystem services to develop a scientific temperament in making collective choices about the ecosystem, by highlighting the direct and indirect uses of ecosystem services for decisive, technical and informational needs (Laurans et al. 2013), the addition of health and non-use values to the economic ones is based on the consideration of these being part of the ecological wealth of a region, which might suffer in time due to the transfer of values to other spheres and degradation due to the pursuit of anthropocentric policies. Together, these values develop better coverage of the ecosystem wealth of a region (Atkinson, Bateman and Mourato 2012), where the diverse nature of use, non-use, ecological and anthropological values could be represented as part of the cultural, provisioning, regulating and supporting functions of ecosystem services (Ojea, Martin-Ortega and Chiabai 2012). Although complex and data-intensive, Gómez-Baggethun and Ruiz-Pérez (2011) believe that ecosystem valuation could relieve policy choices of the arbitrary economic valuation representing commoditized ecosystem services, which would defeat the very purpose of value pluralism and the deliberative judgement process that needs improvement beyond the stated preferences, expert opinions and cost–benefit analyses currently relied upon.

METHODOLOGICAL COMPLEXITIES IN ESTIMATING THE LIFE CYCLE AND EXTERNALITIES OF FA

At the crossroads of economic and scientific understanding, the problem of FA and its externalities is peculiar and complex. While economic and demographic studies can help in evaluating land resources required in developing new FA ash banks or maintaining the existing ones, they could also help in estimating the social willingness-to-pay in avoiding air and water contamination within the region and saving ecological resources. Since FA gets fragmented into its constituent elements, which are also found within benign nature, delayed higher-order impacts are not captured. While prevailing economic methods (like the travel cost method and the stated preferences method) could generate an arbitrary numéraire to estimate the economic burden, this is just arbitrary. As a result, the policy choices in managing waste would lack a complete view of how it might impact the biosphere. This gives rise to the debate of selecting science- versus market-based policy alternatives—market-based choices could be easier, but also reflect a lack of knowledge regarding ecosystem services (Sagoff 2011).

In comparison, cost-based methodologies are dependent on the causal relationships of the involved entities, which follow predetermined relationships. Accordingly, the existing methodologies suffer in cases where causality is replaced by the probabilistic nature of interactions. Translating the problem to the domain of waste management, if waste cannot be traced back to the origin after it has reached the common pool, costing methodologies cannot be applied directly to second- and higher-order impacts and costs for remediation or restoration. While in the available literature, the TCA approach includes multiple perspectives to incorporate hard-to-measure impacts into alternative decisions—some of which has been experimented with as part of the lifecycle design of automobile parts by Carlsson (2007, 2009)—only limited research is available to apply it to the realm of waste management, including a void in physical modelling to represent the behaviour of elements and subsequent impacts.

USE OF DECISION-TREE ANALYSIS TO TRACE THE FLOW OF ENVIRONMENTAL ASPECTS

Due to the shortcomings of the different methods mentioned in the previous section, and their failure to move beyond the first-order impacts of waste, this chapter has used multi-criteria analysis (MCA) to model the waste cycle; its externalities are evaluated using the TCA framework, which can extend to multiple cost chains and include direct, indirect, contingent, hidden and contextual externalities. The literature references MCA and its diverse methodologies for decision-making with regard to the improvement from a one-dimensional approach—that is, reducing the decision variables to the singular dimension of monetary or economic considerations—to consider multiple dimensions such as stakeholders' opinions, business objectives, qualitative gradations, and so on. MCA is used here as the umbrella term to describe a set of approaches that can handle multiple criteria in the decision-making process (Belton and Stewart 2002). MCA can deal with quantitative as well as qualitative data such as ranks, choices and opinions (subjective criteria) of decision makers. Several theories in MCA have extended it to include fuzzy sets to accept decision inputs in a natural language and handle the subjective views of decision makers (Kannan et al. 2013).

A review of the available techniques suggests that MCA can cover three patterns of logic: (a) simple ordering, (b) goal setting or goal seeking and (c) value maximization, including when the criteria or parameters to arrive at the decision might not be well-defined (Gamper, Thöni and Weck-Hannemann 2006). A recent review of MCA by Velasquez and Hester (2013) has compiled a list of popular methods that include multi-attribute utility theory (MAUT), analytical hierarchy process (AHP), fuzzy set theory, case-based reasoning, data envelopment analysis (DEA), simple multi-attribute rating technique (SMART), goal programming (GP), preference ranking organization method for enrichment evaluation (PROMTHEE), elimination and choice expressing reality (ELECTRE), and so on. Interested readers can refer to the standard texts for further reading on these techniques. It has been opined in the literature that MCA can offer methodological refinement beyond CBA and improve decision-making that involves complex problems in the sustainability arena (Bebbington, Brown and Frame 2007; Gamper, Thöni and Weck-Hannemann 2006). For instance, MCA has been explored to model the behaviour of FA upon its disposal, which can be extended to include externalities of material recycling and waste management as part of the decision-making process.

MCA is based on the principle of disaggregating a complex problem into a set of decisions that results in building a decision tree which follows expected utility rules to navigate through the underlying course of action and trace the ultimate outcome of the process, where the likelihood of an outcome is based on the weighted probabilities of individual actions along the chosen path (Barzilai 2010). For a generalized modelling of the waste cycle, its movement can also be traced over different receptors. Accordingly, this would include the impacts of FA and its elements on air, soil, humans, and so on. Based on the flow of the aspects in an environment, their corresponding impacts could be modelled as part of a classification (to accept a qualitative variable: presence/absence or yes/no) and regression tree (if the variable is quantitative). Accordingly, the waste movement is traced till it completes the branching needed to study the desired impact for a select endpoint and can handle aspects across multiple pathways following the real behavior of aspects (Sorvari and Seppälä 2010). Using LCA, a probabilistic flow model can be developed, which can offer some insights into the average level of exposure for different metallic and non-metallic compounds degrading through different levels of receptors. This would offer theoretical grounds to consider the average deposition rate along a pathway. As LCA is a tool to handle physical data, this would not improve economic decision-making process, and necessitates use of the TCA approach to consider potential economic impacts.

TCA TO GENERALIZE EXTERNALITIES OF FA

In the absence of a standardized unit of service to represent ecological considerations impacting any process (Atkinson, Bateman and

Mourato 2012), TCA can leverage a decision-tree model to develop social costs to remediate or mitigate environmental impacts generated by different choices in managing waste. Instead of using a single model to approach the problem, TCA follows multiple approaches to measure different types of impacts. Accordingly, the nature of the costs and their evaluation process could follow different approaches, which would depend on the nature and scope of decision-making. Following from the decision-tree model, costs associated with an impact at an end point i would be sum of the costs associated with the node multiplied by the share of the burden corresponding branches would have to carry. For example, if the actual dispersion of an aspect or element followed multiple pathways, a child branch would bear a proportionate share of the costs. This follows the basic principal of the sum of shared weights at the parent node being equal to one. This linearity in modelling helps in avoiding circular references and overloading a branch. Similar to the objectives in the tree, the flow of aspects is characterized as essential, understandable, operational, non-redundant, concise and preferentially independent (Franco and Montibeller 2009). So, the overall cost function for a policy impact y at an endpoint i will be the sum of costs along the branches C_1, C_2, C_3... with a corresponding share s_1, s_2, s_3...:

$$P(y) = \sum C_i \times s_i \text{ for all } i\text{'s} \quad (A.1)$$

The equation can be further generalized as the total cost function Z:

$$Z = \sum\sum\sum\sum x \times y \times z + \alpha + \beta \quad (A.2)$$

where x = likelihood of the event i due to element j,
$\quad y$ = probability of the event creating an impact k,
$\quad z$ = cost l to remediate/abate impact k due to event i and element j,
$\quad \alpha$ = opportunity losses due to non-recyclability of waste,
and β = regulatory costs.

The next section studies a case from India to explore the fitness of the model.

THEORETICAL EVALUATION OF THE PROPOSED MODEL: A CASE EXAMPLE

This section analyses the operational feasibility of the proposed model. Policy choices can be based on select endpoints and can be mapped by using decision-tree analysis, whereas the imputed cost of externalities can reflect costs to be incurred for remediation, abatement and alternate arrangements within the impact zone. This removes the need to restrict the model to any specific type of cost—say, economic costs—and explores the nature of the problem from multiple angles to enter into meaningful dialogue regarding policy choices. Table A1 reflects the dispersion of FA through different pathways and the corresponding impacts.

Part A: Risk-based cost modeling

For a select endpoint (from Table 1), there could be multiple rounds of cost evaluation involving data sets that could cover each element as

Table A1: *Dispersion of FA aspects in decreasing order of specificity*

Level	Previous level →	Impact
Level 1	--	Fly ash dumped in pond/mound/disposal
Level 2a	Level 1	Surface-to-air dispersion (pathway 1)
Level 2b	Level 1	Surface leaching (pathway 2)
Level 3a	Level 2a	Human respiratory issues due to excess FA exposure (#)
Level 3b	Level 2b	Groundwater contamination
Level 4a	Level 3b	Aquatic/water body contamination
Level 4b	Level 3b	Welfare costs due to consumption of contaminated water (#)
Level 5	Level 4a	Problems to fishes, etc.
Level 6	Level 5	Human consumption of infected fishes (#)
...	...	Higher human/ecological impacts due to further degradability

Note: (#) = endpoints.

part of a given region. The cost formulations would follow, evaluating the total cost of impacts along any particular pathway. The endpoint analysis will, accordingly, help in assessing damages to different ecologic receptors, including humans. For example, site-specific studies can estimate cost of land lost, impact on water resources and health of inhabitants in the nearby areas, which could also include making alternate arrangements to provide water (in case of water contamination), and so on. While an actual on-site study could be one of the ways to formulate and evaluate these and other impacts and generate estimated losses in part, for example, the studies by Rowe, Hopkins and Congdon (2002) and Ruhl et al. (2012) have assessed site-specific damages using different risk factors. However, this type of study would result in an isolated statement of impacts for a given region, where data specificity would reduce the generalizability (Box 1). Another way could be to develop a framework solution that might be representative of each of these parameters such that it could serve as a library of impacts, reflecting the overall condition of ecological damages—for example, the standard risk assessment models developed by EPA (2002, 2010), which are based on the study of risk assessment of different elements of FA for human and ecological receptors, covering a wide array of FA deposition sites in the US.

Once the concentration of an element increases in the ambient environment so that it breaches the ecological response barrier (of countering it), it would increase the risk level, which can be identified through a composite risk number. The composite risk number (corresponding to each element) is dependent on the underlying threat-assessment models that would need to consider multiple factors, such as the ecotoxicological profile of the element, area covered around the source site, geographical and hydrological profile of the area, and various other elements. The aim of this study is not to assess the verifiability or dependability of the model, which would be a study in itself, but to indicate that the comprehensiveness of risk numbers could be a generalized solution to externalities. Accordingly, the cost function Z can be reframed (from equation A.3) to represent the social and environmental costs of chosen policy option as:

$$Z = \sum [(\text{Assessed risk level for } \textit{first element} \text{ at } X^1 \text{ level} \\ \times \text{ social cost to averse the risk at same level}) \\ + (\text{second element}) + (\text{third element}) + \ldots \\ + (\text{nth element})], \text{ for a given pathway-/liner-/ash-} \\ \text{type combination, over all combinations.} \quad (A.3)$$

Box 1: Developing impacts based on specific endpoint analysis

Consider a particular pathway from Table A1, e.g., social welfare costs due to consumption of contaminated water. The flow of aspects will be:

Fly ash dumping → Surface leaching → Underground water contamination → Water body contamination → Ingestion of contaminated water

Endpoints: Humans (direct), other receptors like aquatic life-forms (direct)

The respective models needed to model impacts at each level would be:

a. Surface leaching model explaining the overall flow and rate of contamination due to underground seepage (α with probability of r%),
b. Model predicting contamination of underground water table due to leachating of fly ash (β with probability of s%),
c. Model predicting water-body contamination due to plume/debris deposition by contaminated underground flow (γ with probability of t%), and
d. Human ingestion and physiological characteristics to develop response to specific element(s) of FA (δ with probability of u%).

In other words, the quantity of FA available for contamination

$$= \sum (\delta \times u\%) - u \times (\text{for all elements of FA})$$

Where u^* is the natural level of individual elements available in the biosphere for consumption by receptors. Each of these models would need to provide for regional variations due to geological, geophysical and other region-specific characteristics to be able to predict a benign level of metal and non-mental compounds and a relative measure of risks due to the increase of each of these elements in each of the receptors.

---

[1] X here represents risk assessment level which can differ from average, median, or percentile level, to the site-specific ones. The cost to averse for each level should match it at the same level of specificity.

$$= \sum\sum\sum\sum R_{iplz} \times S_{iplz} \qquad (A.4)$$

where R = risk value for an element, e.g., risk of increased level of arsenic to produce carcinogenic or non-carcinogenic impacts in human receptors,
and S = social costs to abate the risks at the matching level, summed over 'i' elements, 'p' pathways, 'l' liners and 'z' ash type

The formula is recursive and could contain information across elements, pathways (e.g., surface-to-water, groundwater-to-drinking, etc.), liners (no liner, mud liners, polythene liners, etc.) and ash types. To be noted, the cost build-up can add elements like cost of avoidance, cost of abatement and/or restoration costs, and other methodologies (using economic as well as non-economic costing and valuation methods) to arrive at the total cost of any specific damage.

Part B: Replacement resource cost model (to explore opportunity cost of waste)

The opportunity cost of an economic decision reflects the cost of lost opportunity (α from equation A.3) in excess of what is being returned by the prevailing arrangement of resources and is an important parameter in decision-making that represents loss due to the prevailing arrangement of resources. Use of FA in industrial applications represents an opportunity to save the cost of resources that are otherwise being used and should be captured as part of policy choices by using a replacement resource model. For example, FA could be an active ingredient in Portland pozzolana cement (PPC), concrete mix, concrete, bricks, wood-substitute products, soil stabilization, road bases and embankments, land reclamation, and so on, and could save significant monetary and material resources in an economy (Asokan, Saxena and Asolekar 2005; TERI 2006) and can be added to the policy choice.

So, Z' (equivalent cost of lost opportunity)

= [(Waste being used in a productive mix × cost of the final product per unit) × 100/(Total quantity of input mix)]
 − [Cost of transportation of FA + Cost of regulations (if any)] (A.5)

FA AND HIDDEN COSTS

The results from sample evaluation in different cost functions indicate that while industries have claimed ₹150–200 per tonne of FA to be the average cost of its disposal (TERI 2006), an estimate of its externalities works out to be close to ₹1,150 per tonne if considering just two externalities—that is, the cost of cleaning water contamination and average healthcare costs due to burning of coal—indicating that the minimum increase is five- or six-fold of the economic costs (at the present level of its industrial uptake). On the other hand, resource replacement model adds opportunity loss (at ₹211 per tonne of FA) being incurred by not improving its uptake, even when FA remains (mostly) a free-of-cost resource to industries. These estimates can be improved further by bringing in other time-delayed environmental and social improvements as a part of the chain of events. As compared to that, risk-based modelling shows carcinogenic healthcare costs are around ₹2,350 per tonne per year for arsenic contamination alone, assuming median healthcare costs of ₹66 thousand at the 90th percentile for India (which is miniscule compared to the USD 90,000 per cancer patient per year in the US). If we consider exposure cost for a population of 100,000 alone, this would result in ₹14,100 per annum in healthcare costs in fighting carcinogenic arsenic poisoning alone, which is being borne by the social infrastructure today.

With roughly 100 million tonnes of FA getting dumped per annum in ash banks in India and where 400,000 acres of land are already under ash bank, there is an opportunity for the policymakers to work towards policy initiatives to improve the industrial uptake of CCW and fund research efforts to build data banks on its utilization, epidemiological case studies, healthcare impediments, and scientific experiments covering the study of its lab- and site-specific behaviors. Savings could also add ₹41.1 billion per annum of industrial expenditure, currently getting incurred in disposal activities and as part of opportunity losses (estimates vary, but its current industrial uptake is between 20 per cent and 50 per cent in India and elsewhere) (TERI 2006).

DECISION-TREE MODELING AND TRACEABILITY OF HIGHER ORDER IMPACTS

In general, as we move away from first-order impacts, the traceability of second- and higher-order impacts gets diluted and replaced by generalized relations. This progressive generalization of higher-order impacts is time-delayed and limited by our knowledge of ecological complexities. The problem is compounded further due to spatio-temporal variations in the adsorption and dispersion rates of the aspects and their constituent elements, not to mention intra- and inter-generational impacts and the ability of the biosphere to adjust itself to counter the adversaries to maintain balance until a critical level is breached, also pointing to the limits of human cognition. Decision tree-based modelling offers a better view of the causal flow of events and how aspects move through different stage. Instead of using approximations or ignoring the higher-order impacts altogether, a decision tree offers the ability to replicate physical reality, and can be improved over time. By modelling the flow of aspects using a decision tree, the impact of specific policy choices could also be expanded to cover the probabilistic nature of physical realities. Also, in risk-based modelling, the assessment of human and ecological risk numbers would depend on the media concentrations, exposure pathways relevant to a particular medium (EPA 2010), and other factors such as bioaccumulation levels, dose sensitivity and biological reactions of the exposed entity (Rowe, Hopkins and Congdon 2002).

Here, the risk formulation by the EPA (2002, 2010) has been used as the basis for cost computations at a specific percentile level, only for a particular element—arsenic. For example, arsenic has been identified as one of the cancer-causing elements from FA that has a high risk value (EPA 2010) and arsenicosis or elevated As poisoning could be avoided by multiple methods (Pandey et al. 2011); accordingly, the social costs to avert the cancer risk due to arsenic poisoning has been matched to the corresponding risk value at the 90th percentile level. Still, there could be region-specific variations as part of the collected data points.

As mentioned earlier, there is insufficient information available on the standardized abatement costs, cost of avoidance and cost of ecological losses due to these aspects, and accordingly, average values were considered in evaluating the pay-offs along with the probabilities that are yet to be standardized for any region and might not match the risk profiles. This does not take away the fact that lack of data on healthcare costs in the public domain is an issue in ascertaining the preventive or actual treatment costs of malignant diseases like arsenicosis, even where epidemiological studies are sufficient in number—a case in point being arsenic poisoning in Bangladesh and West Bengal (in India), where cost issues have been addressed inadequately (Irfan 2012; Mahmood and Halder 2011; Meij 2003; Smith, Lingas and Rahman 2000; WHO 2000). This indicates the necessity of site-specific studies that would develop representational data to capture parameters and develop quintile-based profiles of risks and costs.

Last but not least, validation of the model for predictability of costs is not the intent of this exploratory research. Rather, it is to open up discussions on a generalizable framework that would use a scientific approach to adapt uncertainties and probabilistic behaviours of aspects and their impacts. The complexity of ecological problems would soon outgrow the capabilities of the existing methods and this would impact the cost of human decisions. A fusion of tools from different disciplines could be one of the ways to improve decision-making further. MCA, TCA and other methods are expected to generate better insights into complex ecological problems and this research was an exploration in that direction.

REFERENCES

Abedil, A., B. Vosough, M. Razani, M. B. Kasiri, S. Steiniger and G. Ebrahimi. 2018. 'Obsidian Deposits from North-western Iran and First Analytical Results: Implications for Prehistoric Production and Trade.' *Mediterranean Archaeology and Archaeometry* 18 (2): 107–118.

Akisik, Oorhan, and Graham Gal. 2011. 'Sustainability in Businesses, Corporate Social Responsibility, and Accounting Standards: An Empirical Study.' *International Journal of Accounting & Information Management* 19 (3): 304–324. https://doi.org/10.1108/18347641111169287.

Aldrich, H., & Auster, E. R. 1986. 'Even Dwarfs Started Small: Liabilities of Age and Size and their Strategic Implications.' *Research in Organizational Behavior*, 8, 165–198.

Albu, Nadia, Catalin Nicolae Albu, Madalina Dumitru and Valentin Florentin Dumitru. 2013. 'Plurality or Convergence in Sustainability Reporting Standards?' *The AMFITEATRU ECONOMIC journal* 15 (special 7): 729–742. Bucharest: Bucharest University of Economic Studies.

Alier, Joan Martinez. 2009. 'Socially Sustainable Economic De-growth.' *Development and Change* 40 (6): 1099–1119. https://doi.org/10.1111/j.1467-7660.2009.01618.x.

Altman, Edward I. 1968. 'Financial Ratios, Discriminant Analysis and the Prediction of Corporate Bankruptcy.' *The Journal of Finance* 23 (4): 589–609. https://doi.org/10.1111/j.1540-6261.1968.tb00843.x.

Alveranga, Tiago HEnrique de Paula, Jairo José Assumpção, Simone Sartori, Lucila Maria de Souza Campos, Mauricio Uriona Maldonado and Fernando Antônio Forcellini. 2013. 'Green Supply Chain Management and Business Process Management: A Union for Sustainable Process in a Furniture Factory.' *Asian Journal of Business & Management Sciences* 4 (2): 1–13. http://www.ajbms.org/articlepdf/1-ajbms02-2015-04-02-4208.pdf.

Antheaume, Nicolas. 2004. 'Valuing External Costs—from Theory to Practice: Implications for Full Cost Environmental Accounting.' *European Accounting Review* 13 (3): 443–464. https://doi.org/10.1080/0963818042000216802.

Anthoff, David, and Richard S. J. Tol. 2010. 'On International Equity Weights and National Decision Making on Climate Change.' *Journal of Environmental Economics and Management* 60 (1): 14–20. https://doi.org/10.1016/j.jeem.2010.04.002.

Aras, Güler, and David Crowther. 2008. 'Governance and Sustainability: An Investigation into the Relationship between Corporate Governance and Corporate Sustainability.' *Management Decision* 46 (3): 433–448. https://doi.org/10.1108/00251740810863870.

Arivazhagan, K., M. Ravichanrdan, S. K. Dube, V. K. Mathur, Ram Krishna Khandakar, M. M. Kamal Pasha, A. K. Sinha, B. D. Sarangi, V. K. M. Tripathi, S. K. Gupta, Rajvir Singh, Mushtaq Ali, A. S. Thakur and Raghvendra Narayan. 2011. 'Effect of Coal Fly Ash on Agricultural Corps: Showcase Project on Use of Fly Ash in Agriculture in and around Thermal Power Station Areas of National Thermal Power Corporation Ltd., India.' Paper presented at the World of Coal Ash Conference, 9–12 May 2011, Denver, Colorado, USA.

Asokan, P., Mohini Saxena and Shyam R. Asolekar. 2005. 'Coal Combustion Residues—Environmental Implications and Recycling Potentials.' *Resources, Conservation and Recycling* 43 (3): 239–262. https://doi.org/10.1016/j.resconrec.2004.06.003.

Atkinson, Giles, Ian Bateman and Susana Mourato. 2012. 'Recent Advances in the Valuation of Ecosystem Services and Biodiversity.' *Oxford Review of Economic Policy* 28 (1): 22–47. https://doi.org/10.1093/oxrep/grs007.

Atkinson, Giles. 2008. 'Sustainability, the Capital Approach and the Built Environment.' *Building Research & Information* 36 (3): 241–247. https://doi.org/10.1080/09613210801900734.

Auster, Ellen, and Howard E. Aldrich. 1984. 'Small Business Vulnerability, Ethnic Enclaves, and Ethnic Enterprise.' In *Ethnic Communities in Business: Strategies for Economic Survival*, edited by Robin Ward and Richard Jenkins, 39–54. Cambridge/New York: Cambridge University Press.

Bainbridge, David A. 2007. 'True Cost Accounting for a Post-Autistic Economy.' *Post-Autistic Economic Review* 41: 23–28. https://rppe.files.wordpress.com/2007/03/paer-issue-no-41.pdf.

Baladouni, Vahé. 1984. 'Etymological Observations on Some Accounting Terms.' *The Accounting Historians Journal* 11 (2): 101–109. https://doi.org/10.2308/0148-4184.11.2.101.

Baldvinsdottir, Gudrun, Falconer Mitchell and Hanne Nørreklit. 2010. 'Issues in the Relationship between Theory and Practice in Management Accounting.' *Management Accounting Research* 21 (2): 79–82. https://doi.org/10.1016/j.mar.2010.02.006.

Ball, Ray, and Philip Brown. 1968. 'An Empirical Evaluation of Accounting Income Numbers.' *Journal of Accounting Research* 6 (2): 159–178. https://doi.org/10.2307/2490232.

Barbier, Edward B. 2013. 'Wealth Accounting, Ecological Capital and Ecosystem Services.' *Environment and Development Economics* 18 (2): 133–161. Https://doi.org/10.1017/S1355770X12000551.

Barzilai, Jonathan. 2010. 'Preference Function Modelling: The Mathematical Foundations of Decision Theory.' In *Trends in Multiple Criteria Decision Analysis*, edited by Matthias Ehrgott M, José Rui Figueira and Salvatore Greco, 57–86. New York: Springer. https://doi.org/10.1007/978-1-4419-5904-1.

Bates, Timothy, and Alfred Nucci. 1989. 'An Analysis of Small Business Size and Rate of Discontinuance. *Journal of Small Business Management*, 27 (4): 1–8.

Baunsgaard, Vibeke Vad, and Stewart Clegg. 2015. 'Innovation: A Critical Assessment of the Concept and Scope of Literature.' In *The Handbook of Service Innovation*, edited by Renu Agarwal, Willem Selen, Göran Roos and Roy Green, 5–25. London: Springer. https://doi.org/10.1007/978-1-4471-6590-3.

Beaver, William H. 1966. 'Financial Ratios as Predictors of Failure.' *Journal of Accounting Research* 4: 71–111. https://doi.org/10.2307/2490171.

Beaver, William H., Maureen F. McNichols and Jung-Wu Rhie. 2005. 'Have Financial Statements Become Less Informative? Evidence from the Ability of Financial Ratios to Predict Bankruptcy.' *Review of Accounting Studies* 10 (1): 93–122. https://doi.org/10.1007/s11142-004-6341-9.

Bebbington, Jan, and Carlos Larrinaga-González. 2008. 'Carbon Trading: Accounting and Reporting Issues.' *European Accounting Review* 17 (4): 697–717. https://doi.org/10.1080/09638180802489162.

Bebbington, Jan, Judy Brown and Bob Frame. 2007. 'Accounting Technologies and Sustainability Assessment Models.' *Ecological Economics* 61 (2–3): 224–236. https://doi.org/10.1016/j.ecolecon.2006.10.021.

Bebbington, Jan, Rob Gray, Chris Hibbitt and Elizabeth A. Kirk. 2001. *Full Cost Accounting: An Agenda for Action*. London, UK: Certified Accountants Educational Trust.

Beck, Cornelia, John Dumay and Geoffrey Frost. 2017. 'In Pursuit of a "Single Source of Truth": From Threatened Legitimacy to Integrated Reporting.' *Journal of Busness Ethics* 141 (1): 191–205. https://doi.org/10.1007/s10551-014-2423-1.

Belton, Valerie, and Theodor J. Stewart. 2002. *Multiple Criteria Decision Analysis: An Integrated Approach*. Massachusetts: Kluwer Academic Publishers. https://doi.org/10.1007/978-1-4615-1495-4.

Birkin, Frank, and Thomas Polesie. 2011. 'An Epistemic Analysis of (Un)Sustainable Business.' *Journal of Business Ethics* 103 (2): 239–253. https://doi.org/10.1007/s10551-011-0863-4.

Block, Walter. 1998. 'Environmentalism and Economic Freedom: The Case for Private Property Rights.' *Journal of Business Ethics* 17 (16): 1887–1899.

Böhringer, Christoph, and Patrick E. P. Jochem. 2007. 'Measuring the Immeasurable—A Survey of Sustainability Indices.' *Ecological Economics* 63 (1): 1–8. https://doi.org/10.1016/j.ecolecon.2007.03.008.

Boone, Christophe A. J. J., and Arjen van Witteloostuijn. 1995. 'Industrial Organization and Organizational Ecology: The Potentials for Cross-fertilization.' *Organization Studies* 16 (2): 265–298. https://doi.org/10.1177/017084069501600204.

Bose, Anu, and Ian Blore. 1993. 'Public Waste and Private Property: An Enquiry into the Economics of Solid Waste in Calcutta.' *Public Administration and Development* 13 (1): 1–15. https://doi.org/10.1002/pad.4230130102.

Boström, Magnus, and Mikael Klintman. 2008. *Eco-Standards, Product Labelling and Green Consumerism*. Basingstoke: Palgrave Macmillan.

Bracci, Enrico, and Laura Maran. 2013. 'Environmental Management and Regulation: Pitfalls of Environmental Accounting?' *Management of Environmental Quality: An International Journal* 24 (4): 538–554. https://doi.org/10.1108/MEQ-04-2012-0027.

Branco, Manuel Castelo, and Lucia Lima Rodrigues. 2007. 'Issues in Corporate Social and Environmental Reporting Research: An Overview.' *Issues in Social and Environmental Accounting* 1 (1): 72–90. https://doi.org/10.22164/isea.v1i1.9.

Braungart, Michael, William McDonough and Andrew Bollinger. 2007. 'Cradle-to-Cradle Design: Creating Healthy Emissions as a Strategy for Eco-effective Product and System Design.' *Journal of Cleaner Production* 15 (13–14): 1337–1348. https://doi.org/10.1016/j.jclepro.2006.08.003.

Brennan, Niamh M., and Doris M. Merkl-Davies. 2014. 'Rhetoric and Argument in Social and Environmental Reporting: The Dirty Laundry Case.' *Accounting, Auditing & Accountability Journal* 27 (4): 602–633. https://doi.org/10.1108/AAAJ-04-2013-1333.

Broughton, Emma, and Romain Pirard. 2011. 'What's in a Name? Market-based Instruments for Biodiversity.' In *Health and Environmental Reports* 8: 1–44.

Brown, Judy, and Michael Fraser. 2006. 'Approaches and Perspectives in Social and Environmental Accounting: An Overview of the Conceptual Landscape.' *Business Strategy and the Environment* 15 (2): 103–117. https://doi.org/10.1002/bse.452.

Brown, Judy, and Jesse Dillard. 2014. 'Integrated Reporting: On the Need for Broadening Out and Opening Up.' *Accounting, Auditing & Accountability Journal* 27 (7): 1120–1156. https://doi.org/10.1108/AAAJ-04-2013-1313.

Brown, Mark T., and Sergio Ulgiati. 2004. 'Energy Quality, Emergy, and Transformity: H. T. Odum's Contributions to Quantifying and Understanding Systems.' *Ecological Modelling* 178 (1–2): 201–213.

Brüderl, Josef, Peter Preisendörfer and Rolf Ziegler. 1992. 'Survival Chances of Newly Founded Business Organizations.' *American Sociological Review* 57 (2): 227–242. https://doi.org/10.2307/2096207.

Burritt, Roger L., Christian Herzig and Bernardo D. Tadeo. 2009. 'Environmental Management Accounting for Cleaner Production: The Case of a Philippine Rice Mill.' *Journal of Cleaner Production* 17 (4): 431–439. https://doi.org/10.1016/j.jclepro.2008.07.005.

Burritt, Roger L., Stefan Schaltegger and Dimitar Zvezdov. 2011. 'Carbon Management Accounting: Explaining Practice in Leading German Companies.' *Australian Accounting Review* 21 (1): 80–98. https://doi.org/10.1111/j.1835-2561.2010.00121.x.

Carlsson, Bo T. 2007. 'Suitability Analysis of Selective Solar Absorber Surfaces Based on a Total Cost Accounting Approach.' *Solar Energy Materials & Solar Cells* 91 (14): 1338–1349. https://doi.org/10.1016/j.solmat.2007.05.011.

Carlsson, Bo T. 2009. 'Selecting Material for the Exterior Panel of a Private Car Back Door by Adopting a Total Cost Accounting Approach.' *Materials and Design* 30 (3): 826–832. https://doi.org/10.1016/j.matdes.2008.05.031.

Carter, Robert M., Chris R. de Freitas, Indur M. Goklany, David Holland, Richard S. Lindzen, Ian Byatt, Ian Castles, David Henderson, Nigel Lawson, Ross McKitrick, Julian Morris, Alan Peacock, Colin Robinson and Robert Skidelsky. 2007. 'The Stern Review: A Dual Critique.' *World Economics* 7 (4): 165–232.

Célérier, Laure, and Luis Emilio Cuenca Botey. 2015. 'Participatory Budgeting at a Community Level in Porto Alegre: A Bourdieusian Interpretation.' *Accounting, Auditing & Accountability Journal* 28 (5): 739–772. https://doi.org/10.1108/AAAJ-03-2013-1245.

Central Electricity Authority. 2011. CO_2 *Baseline Database for the Indian Power Sector: User Guide, version 6.0.* Retrieved from http://www.cea.nic.in/reports/planning/cdm_co2/user_guide_ver6.pdf (last accessed 15 January 2017).

Chaturvedi, Bharati. 2003. 'Waste-handlers and Recycling in Urban India: Policy, Perception and the Law.' *Social Change* 33 (2–3): 41–50. https://doi.org/10.1177/004908570303300304.

Coase, Ronald H. 1937. 'The Nature of the Firm.' *Economica* 4 (16): 386–405. https://doi.org/10.1111/j.1468-0335.1937.tb00002.x.

Comite, Ubaldo. 2009. 'The Evolution of a Modern Business from Its Assets and Liabilities Statement to Its Ethical Environmental Account.' *Journal of Management Research* 9 (2): 100–120.

Connolly, Ciaran J., and Noel S. Hyndman. 2004. 'Performance Reporting: A Comparative Study of British and Irish Charities.' *The British Accounting Review* 36 (2): 127–154. https://doi.org/ 10.1016/j.bar.2003.10.004.

Cowen, T. K. 1968. 'A Pragmatic Approach to Accounting Theory.' *The Accounting Review* 43 (1): 94–100.

Curkovic, Sime, and Robert Sroufe. 2007. 'Total Quality Environmental Management and Total Cost Assessment: An Exploratory Study.' *International Journal of Production Economics* 105 (2): 560–579. https://doi.org/10.1016/j.ijpe.2006.04.021.

Cyert, Richard M., and James G. March. 1963. *A Behavioral Theory of the Firm.* University of Illinois at Urbana-Champaign's Academy for Entrepreneurial Leadership Historical Research Reference in Entrepreneurship.

Dahl, Arthur Lyon. 2012. 'Achievements and Gaps in Indicators for Sustainability.' *Ecological Indicators* 17 (June): 14–19. https://doi.org/10.1016/j.ecolind.2011.04.032.

Dahlman, Carl J. 1979. 'The Problem of Externality.' *The Journal of Law & Economics* 22 (1): 141–162. https://doi.org/10.1086/466936.

Daly, Herman E. 2005. 'Economics in a Full World.' *Scientific American*, September 2005. https://www.scientificamerican.com/article/economics-in-a-full-world.

Dascalu, Cornelia, Chirata Caraiani, Camelia Iuliana Lungu, Florian Colceag and Gina Raluca Guse. 2010. 'The Externalities in Social Environmental Accounting.' *International Journal of Accounting & Information Management* 18 (1): 19–30. https://doi.org/10.1108/18347641011023252.

Dawkins, Cedric E., and John W. Fraas. 2011. 'Erratum to: Beyond Acclamations and Excuses: Environmental Performance, Voluntary Environmental Disclosure and the Role of Visibility.' *Journal of Business Ethics* 99 (3): 383–397. https://doi.org/10.1007/s10551-010-0659-y.

Debnath, S. 2015. 'Integrated Waste Management Framework: A Business Case from Hospitality Industry.' *International Journal of Business Excellence* 8 (5): 566–583. https://doi.org/10.1504/IJBEX.2015.071278.

Debnath, S. 2016. 'Implementing Environmental Management Accounting (EMA): A Case Study from India.' In *Corporations and Sustainability: The South Asian Perspective*, edited by P. D. Jose, 34–54. London: Routledge.

Debnath, S., and S. K. Bose. 2014. 'Exploring Full Cost Accounting Approach to Evaluate Cost of MSW Services in India.' *Resources, Conservation and Recycling* 83: 87–95. https://doi.org/10.1016/j.resconrec.2013.12.007.

Debnath, Somnath. 2014. 'Expanding Environmental Management Accounting: An Experimental Construct to Integrate Material Wastes and Emission Flows.' *International Journal of Business Information Systems* 16 (2): 119–133. https://doi.org/10.1504/IJBIS.2014.062834.

Dorbian, Christopher S., Philip J. Wolfe and Ian A. Waitz. 2011. 'Estimating the Climate and Air Quality Benefits of Aviation Fuel and Emissions Reductions.' *Atmospheric Environment* 45 (16): 2750–2759. https://doi.org/10.1016/j.atmosenv.2011.02.025.

Dorweiler, Vernon P., and Mehenna Yakhou. 2005. 'A Perspective on the Environment's Balance Sheet.' *The Journal of American Academy of Business, Cambridge* 7 (2): 16–22.

Dumitru, Madalina, and Raluca Gina Guşe. 2017. 'The Legitimacy of the International Integrated Reporting Council.' *Journal of Accounting and Management Information Systems* 16 (1): 30–58. https://doi.org/10.24818/jamis.2017.01002.

Dyckman, Thomas R. 2016. 'Significance Testing: We Can Do Better.' *Abacus* 52 (2): 319–342. https://doi.org/10.1111/abac.12078.

Edwards, John Richard, and Edmund Newell. 1991. 'The Development of Industrial Costs and Management Accounting before 1850: A Survey of the Evidence.' *Business History* 33 (1): 35–57. https://doi.org/10.1080/00076799100000003.

Ehrenfeld, John R. 2005. 'Eco-efficiency: Philosophy, Theory, and Tools.' *Journal of Industrial Ecology* 9 (4): 6–8.

Eisenhardt, Kathleen M. 1989. 'Building Theories from Case Study Research.' *Academy of Management Review* 14 (4): 532–550. https://doi.org/10.2307/258557.

EPA (Environmental Protection Agency). 1997. *Full Cost Accounting for Municipal Soild Waste Management: A Handbook*. Washington, DC: Office of Solid Waste and Emergency Response, United States Environmental Protection Agency (US EPA). https://permanent.access.gpo.gov/lps43186/fca-hanb.pdf

EPA. 2002. 'Constituent Screening for Coal Combustion Wastes.' Draft report, Research Triangle Institute for Office of Solid Waste, Washington, DC.

EPA. 2010. 'Human and Ecological Risk Assessment of Coal Combustion Wastes.' Draft report, EPA.

Epstein, Paul R., Jonathan J. Buonocore, Kevin Eckerle, Michael Hendryx, Benjamin M. Stout III, Richard Heinberg, Richard W. Clapp, Beverly May, Nancy L. Reinhart, Melissa M. Ahern, Samir K. Doshi and Leslie Glustrom. 2011. 'Full Cost Accounting for the Life Cycle of Coal.' In *Ecological Economics Reviwes*, edited by Robert Costanza, Karin Limburg and Ida Kubiszewski, *Annals of the New York Academy of Sciences* 1219 (1): 73–98. https://doi.org/10.1111/j.1749-6632.2010.05890.x.

Esty, Daniel C., and Andrew S. Winston. 2006. *Green to Gold: How Smart Companies Use Environmental Strategy to Innovate, Create Value, and Build Competitive Advantage*. Hoboken, New Jersey: John Wiley & Sons, Inc.

Eugénio, Teresa, Isabel Costa Lourenço and Ana Isabel Morais. 2010. 'Recent Developments in Social and Environmental Accounting Research.' *Social Responsibility Journal* 6 (2): 286–305. https://doi.org/10.1108/17471111011051775.

Firoz, Mohammad, and A. Aziz Ansari. 2010. 'Environmental Accounting and International Financial Reporting Standards (IFRS).' *International Journal of Business and Management* 5 (10): 105–112. https://doi.org/10.5539/ijbm.v5n10p105.

Fisher, Irving. 1930. 'The Economics of Accountancy.' *The American Economic Review* 20 (4): 603–618.

Flinders, C. L. (2007). 'Beyond Male and Female: Mystics Forged Paths that Transcend Gender.' *Science & Spirit*, 18(4), 42–44.

Franco, Luis A., and Gilberto Montibeller, G. 2009. *Problem Structuring for Multicriteria Decision Analysis Interventions*. London, UK: London School of Economics.

Frankham, Richard, Jonathan D. Ballou, Michele R. Dudash, Mark D. B. Eldridge, Charles B. Fenster, Robert C. Lacy, Joseph R. Mendelson III, Ingrid J. Porton, Katherine Ralls and Oliver A. Ryder. 2012. 'Implications of Different Species Concepts for Conserving Biodiversity.' *Biological Conservation* 153: 25–31. https://doi.org/10.1016/j.biocon.2012.04.034.

Fraser, Michael. 2012. '"Fleshing Out" an Engagement with a Social Accounting Technology.' *Accounting, Auditing & Accountability Journal* 25 (3): 508–534. https://doi.org/10.1108/09513571211209626.

Frederiksen, Claus Strue, and Morten Ebbe Juul Nielsen. 2013. 'The Ethical Foundations for CSR.' In *Corporate Social Responsibility: Challenges, Opportunities and Strategies for 21st-Century Leaders*, edited by John O. Okpara and Samuel O. Idowu, 17–33. Berlin, Heidelberg: Springer. https://doi.org/10.1007/978-3-642-40975-2_2.

Freeman, R. Edward. 1984. *Strategic Management: A Stakeholder Approach*. Boston: Pitman

Friedman, M. 1970. 'The Social Responsibility of a Business is to Increase its Profits.' *The New York Times magazine*, 13 September 1970.

Funtowicz, Silvio O., and Jerome R. Ravetz. 1994. 'The Worth of a Songbird: Ecological Economics as a Post-normal Science.' *Ecological Economics*, 10 (3): 197–207. https://doi.org/10.1016/0921-8009(94)90108-2.

Gadenne, David, and Monir Zaman. 2002. 'Strategic Environmental Management Accounting: An Exploratory Study of Current Corporate Practice and Strategic Intent.' *Journal of Environmental Assessment Policy and Management* 4 (2): 123–150. https://doi.org/10.1142/S1464333202000954.

Gale, Robert. 2006. 'Environmental Costs at a Canadian Paper Mill: A Case Study of Environmental Management Accounting (EMA).' *Journal of Cleaner Production* 14 (14): 1237–1251. https://doi.org/10.1016/j.jclepro.2005.08.010.

Gamper, C. D., M. Thöni and H. Weck-Hannemann. 2006. 'A Conceptual Approach to the Use of Cost Benefit and Multi Criteria Analysis in Natural Hazard Management.' *Natural Hazards & Earth System Sciences* 6 (2): 293–302. https://doi.org/10.5194/nhess-6-293-2006.

Garza, MAría Dolores, Albino Prada, Manuel Varela, María Xosé Vázquez Rodríguez. 2009. 'Indirect Assessment of Economic Damages from the Prestige Oil Spill: Consequences for Liability and Risk Prevention.' *Disasters* 33 (1): 95–109. https://doi.org/10.1111/j.0361-3666.2008.01064.x.

Gayer, Ted, and John K. Horowitz. 2005. 'Market-based Approaches to Environmental Regulation.' *Foundations and Trends in Microeconomics* 1 (4): 201–326. https://doi.org/10.1561/0700000013.

Geissdoerfer, Klaus, Ronald Gleich, Andreas Wald and Jaideep Motwani. 2009. 'State and Development of Life-cycle Cost Analysis Models in Strategic Cost Management.' *Production and Inventory Management Journal* 45 (1): 66–79.

Global Reporting Initiative (GRI). 2006. *Indicator Protocol Set—Environment*. Amsterdam, The Netherlands: Global Reporting Initiative. Retrieved from https://www.globalreporting.org/standards/gri-standards-download-center/

Gluch, Pernilla, and Henrikke Baumann. 2004. 'The Life Cycle Costing (LCC) Approach: A Conceptual Discussion of its Usefulness for Environmental Decision-making.' *Building and Environment* 39 (5): 571–580. https://doi.org/10.1016/j.buildenv.2003.10.008.

Goldratt, Elihayu M., and Jeff Cox. 2016. *The Goal: A Process of Ongoing Improvement*. London, UK: Routledge.

Gómez-Baggethun, Erik, and Manuel Ruiz-Pérez. 2011. 'Economic Valuation and the Commodification of Ecosystem Services.' *Progress in Physical*

Geography: Earth and Environment 35 (5): 613–628. https://doi.org/10.1177/0309133311421708.

Gowdy, John, Charles Hall, Kent Klitgaard and Lisi Krall. 2010. 'What Every Conservation Biologist Should Know about Economic Theory.' *Conservation Biology* 24 (6): 1440–1447. https://doi.org/10.1111/j.1523-1739.2010.01563.x.

Gradel, Thomas E., and Braden R. Allenby. 2003. *Industrial Ecology* (2nd edn). Englewood Cliffs, NJ: Prentice Hall.

Granlund, Markus, and Teemu Malmi. 2002. 'Moderate Impact of ERPS on Management Accounting: A Lag or Permanent Outcome?' *Management Accounting Research* 13 (3): 299–321. https://doi.org/10.1006/mare.2002.0189.

Gray, R. 2008. 'Social and Environmental Accounting and Reporting: From Ridicule to Revolution? From Hope to Hubris?—A Personal Review of the Field.' *Issues in Social and Environmental Accounting* 2 (1): 3–18. https://doi.org/10.22164/isea.v2i1.22.

Gray, R., and Jan Bebbington. 2001. *Accounting for the Environment* (2nd ed.). London: SAGE Publications.

Gray, R., and Richard Laughlin. 2012. 'It Was 20 Years Ago Today: Sgt Pepper.' *Accounting, Auditing & Accountability Journal: Green Accounting and the Blue Meanies* 25 (2): 228–255. https://doi.org/10.1108/09513571211198755.

Gray, Rob. 2006. 'Social, Environmental and Sustainability Reporting and Organisational Value Creation: Whose Value? Whose Creation?' *Accounting, Auditing & Accountability Journal* 19 (6): 793–819. https://doi.org/10.1108/09513570610709872

Hájek, Miroslav, Karel Pulkrab and Jaroslava Hyršlová. 2012. 'Forestry Externalities in the Environmental Management Accounting System.' *Problems of Management in the 21st Century* 5 (1): 31–45. http://oaji.net/articles/2014/450-1391966200.pdf.

Haji, Abdifatah Ahmed, and Dewan Mahboob Hossain. 2016. 'Exploring the Implications of Integrated Reporting on Organisational Reporting Practice: Evidence from Highly Regarded Integrated Reporters.' *Qualitative Research in Accounting & Management* 13 (4): 415–444. https://doi.org/10.1108/QRAM-07-2015-0065.

Halton, Celia. 2015. 'Under the Dome: The Smog Film Taking China by Storm.' BBC News, 2 March 2015. China Blog. https://www.bbc.com/news/blogs-china-blog-31689232.

Hannan, Michael T., and John Freeman. 1984. 'Structural Inertia and Organizational Change.' *American Sociological Review* 49 (2): 149–164. https://doi.org/10.2307/2095567.

Hazelton, James. 2013. 'Accounting as a Human Right: The Case of Water Information.' *Accounting, Auditing & Accountability Journal* 26 (2): 267–311. https://doi.org/10.1108/09513571311303738.

Herbohn, Kathleen. 2005. 'A Full Cost Environmental Accounting Experiment.' *Accounting, Organizations and Society* 30 (6): 519–536. HTTPS://doi.org/10.1016/j.aos.2005.01.001.

Hopkinson, Peter, Anthony Sammut and Michael Whitaker. 1999. 'The Standardization of Environmental Performance Indicators and their Relationship to Corporate Environmental Reporting: What Can We Learn from the UK Water Industry?' *Journal of Environmental Assessment Policy and Management* 1 (3): 277–296. https://doi.org/10.1142/S1464333299000235.

Hopwood, A. G. 2009. 'The Economic Crisis and Accounting: Implications for the Research Community.' *Accounting, Organizations and Society* 34 (6–7): 797–802. http://dx.doi.org/10.1016/j.aos.2009.07.004.

Hopwood, A. G. 2008. 'Changing Pressures on the Research Process: On Trying to Research in an Age when Curiosity is not Enough.' *European Accounting Review* 17 (1): 87–96. https://doi.org/10.1080/09638180701819998.

Hopwood, Anthony G. 2007. 'Whither Accounting Research?' *The Accounting Review* 82 (5): 1365–1374. https://doi.org/10.2308/accr.2007.82.5.1365.

Horngren, C. T., Srikant M. Datar, George Foster, Madhav V. Rajan and Chris M. Ittner. 2009. *Cost Accounting: A Managerial Emphasis* (13th edn). New Delhi: Pearson Education.

Horngren, Charles T. 2004. 'Management Accounting: Some Comments.' *Journal of Management Accounting Research* 16 (1): 207–211. https://doi.org/10.2308/jmar.2004.16.1.207.

Hoskin, Keith W., and Richard H. Macve. 2000. 'Knowing More as Knowing Less? Alternative Histories of Cost and Management Accounting in the U.S. and the U.K.' *Accounting Historians Journal* 27 (1): 91–149. http://umiss.lib.olemiss.edu:82/record=b1032496.

Huang, Cheng-Li, and Fan-Hua Kung. 2010. 'Drivers of Environmental Disclosure and Stakeholder Expectation: Evidence from Taiwan.' *Journal of Business Ethics* 96 (3): 435–451. https://doi.org/10.1007/s10551-010-0476-3.

Huang, Y. Anny, Manfred Lenzen, Christopher L. Weber, Joy Murray and H. Scott Matthews. 2009. 'The Role of Input–Output Analysis for Screening of Carbon Footprints.' *Economic Systems Research* 21 (3): 217–242. https://doi.org/10.1080/09535310903541348.

Hyrslová, Jaroslava, and Miroslav Hájek. 2005. 'Environmental Management Accounting in the Framework of EMAS II in the Czech Republic.' In *Implementing Environmental Management Accounting: Status and Challenges—Eco-Efficiency in Industry and Science*, vol. 18, edited by Pall M. Rikhardsson, Martin Bennett, Jan Jaap Bouma and Stefan Schaltegger, 279–295. Dordrecht, The Netherlands: Springer. https://doi.org/10.1007/1-4020-3373-7_14.

IATA. 2008. Aviation Carbon Offset Programmes—IATA Guidelines and Toolkit. Retrieved from https://www.iata.org/

Ienciu, Ionel A. (2009). 'Environmental Performance versus Economic Performance.' *International Journal of Business Research* 9 (5): 125–131.

Ijiri, Yuji. 1987. 'Three Postulates of Momentum Accounting.' *Accounting Horizons* 1 (1): 25–34.

International Federation of Accountants (IFAC). 2005. *International Guidance Document: Environmental Management Accounting*. New York, US: International

Federation of Accountants. https://www.ifac.org/publications-resources/international-guidance-document-environmental-management-accounting.

International Financial Reporting Standards (IFRS) Foundation. 2018. 'Why Global Accounting Standards?' https://www.ifrs.org/use-around-the-world/why-global-accounting-standards.

International Organization for Standardization (ISO). 2011. ISO 14051:2011: 'Environmental Management—Material Flow Cost Accounting—General Framework.' Published September 2011; last confirmed 2018. https://www.iso.org/standard/50986.html. International Rivers. 2016. 'Questions and Answers About Large Dams.' https://www.internationalrivers.org/questions-and-answers-about-large-dams

Irfan, Zareena Begum. 2012. 'Arsenic Contamination in Water: A Conceptual Framework of Policy Options.' Working paper 64/2012, Madras School of Economics, Chennai.

Jacobs, Benjamin. 2011. 'Debris Mitigation Certification and the Commercial Space Industry: A New Weapon in the Fight against Space Pollution.' *Media Law & Policy* 20 (1): 117–141.

James, Marianne L. 2015. 'Accounting Majors' Perceptions of the Advantages and Disadvantages of Sustainability and Integrated Reporting.' *Journal of Legal, Ethical and Regulatory Issues* 18 (2): 107–123.

Jasch, C. M. 2006. 'How to Perform an Environmental Management Cost Assessment in One Day.' *Journal of Cleaner Production* 14 (14): 1194–1213. https://doi.org/10.1016/j.jclepro.2005.08.005.

Jasch, C. M., and Alexander Lavicka. 2006. 'Pilot Project on Sustainability Management Accounting with the Styrian Automobile Cluster.' *Journal of Cleaner Production* 14: 1214–1227. https://doi.org/10.1016/j.jclepro.2005.08.007.

Jasch, C. M., and Myrtille Danse. 2005. 'Environmental Management Accounting Pilot Projects in Costa Rica.' In *Implementing Environmental Management Accounting: Status and Challenges*, edited by Pall M. Rikhardsson, Martin Bennett, Jan Jaap Bouma and Stefan Schaltegger, 343–364. Dordrecht, The Netherlands: Springer. https://doi.org/10.1007/1-4020-3373-7_17.

Jasch, C. M., Daniel Ayres and Ludovic Bernaudat. 2010. 'Environmental Management Accounting (EMA) Case Studies in Honduras—An Integrated UNIDO Project.' *Issues in Social and Environmental Accounting* 4 (2): 89–103. https://doi.org/10.22164/isea.v4i2.48.

Jasch, Christine Maria. 2003. 'The Use of Environmental Management Accounting (EMA) for Identifying Environmental Costs.' *Journal of Cleaner Production* 11 (6): 667–676. https://doi.org/10.1016/S0959-6526(02)00107-5.

Johnson, H. T., and Kaplan, R. S. 1987. 'The Rise and Fall of Management Accounting [2]'. Strategic Finance, 68(7): 22.

Jollands, Nigel, Jonathan Lermit and Murray Patterson. 2003. 'The Usefulness of Aggregate Indicators in Policy Making and Evaluation: A Discussion with

Application to Eco-Efficiency Indicators in New Zealand.' Working/technical paper. https://openresearch-repository.anu.edu.au/handle/1885/41033.

Jones, Michale John, and Jill Frances Solomon. 2013. 'Problematising Accounting for Biodiversity.' *Accounting, Auditing & Accountability Journal* 26 (5): 668–687. https://doi.org/10.1108/AAAJ-03-2013-1255.

Kahya, Emel, and Panayiotis Theodossiou. 1999. 'Predicting Corporate Financial Distress: A Time-Series CUSUM Methodology.' *Review of Quantitative Finance and Accounting* 13 (4): 323–345. https://doi.org/10.1023/A:1008326706404.

Kajikawa, Yuya, Toshihiro Inoue and Thong Ngee Goh. 2011. 'Analysis of Building Environment Assessment Frameworks and their Implications for Sustainability Indicators.' *Sustainability Sciences* 6 (2): 233–246. https://doi.org/10.1007/s11625-011-0131-7.

Kakkuri-Knuuttila, Maria-Liisa, Kari Lukka and Jaakko Kuorikoski. 2008. 'Straddling between Paradigms: A Naturalistic Philosophical Case Study on Interpretive Research in Management Accounting.' *Accounting, Organizations and Society* 33 (2–3): 267–291. https://doi.org/10.1016/j.aos.2006.12.003.

Kannan, Devika, Roohollah Khodarverdi, Laya Olfat, Ahmad Jafarian and Ali Diabat. 2013. 'Integrated Fuzzy Multi Criteria Decision Making Method and Multi-Objective Programming Approach for Supplier Selection and Order Allocation in a Green Supply Chain.' *Journal of Cleaner Production* 47: 355–367. https://doi.org/10.1016/j.jclepro.2013.02.010.

Kaplan, Robert S., and Anthony A. Atkinson. 2009. *Advanced Management Accounting* (13th edition). New Delhi, India: PHI Learning Pvt. Ltd.

Kasai, J. 1999. 'Life Cycle Assessment, Evaluation Method for Sustainable Development.' *JSAE (Journal of Society of Automotive Engineers) Review* 20 (3): 387–393. http://dx.doi.org/10.1016/S0389-4304(99)00013-2.King, Andrew A., and Michawl J. Lenox. 2001. 'Does it Really Pay to be Green? An Empirical Study of Firm Environmental and Financial Performance.' *Journal of Industrial Ecology* 5 (1): 105–116. https://doi.org/10.1162/108819801753358526.

Kitada, Hirotsugu, Hitoshi Okada and Katsuhiko Kokubu. 2009. 'Material Flow Cost Accounting at Japanese Medium Size Company.' *Collected Papers of CSEAR 2009—Christchurch: 8th Australasian Conference on Social and Environment Accounting Research*, edited by Thomas Kern, Nicholas McGuigan, Amanda Ball and Markus J. Milne, 1–16. Christchurch, New Zealand: CSEAR 2009. https://www.researchgate.net/publication/267401663.

Klein, N. 2007. *The Shock Doctrine: The Rise of Disaster Capitalism*. New York, NY, US: Metropolitan Books/Henry Holt and Company.

Kokubu, K., and Michiyasu Nakajima. 2004. 'Material Flow Cost Accounting in Japan: A New Trend of Environmental Management Accounting Practices.' Paper presented at the Fourth Asia Pacific Interdisciplinary Research in Accounting Conference, 4–6 July 2004, Singapore. https://www.researchgate.net/publication/267401663_MATERIAL_FLOW_COST_ACCOUNTING_

IN_JAPAN_A_NEW_TREND_OF_ENVIRONMENTAL_MANAGEMENT_ ACCOUNTING_PRACTICES.

Kokubu, K., and Eriko Nashioka. 2005. 'Environmental Management Accounting Practices in Japan.' In *Implementing Environmental Management Accounting: Status and Challenges*, edited by Pall M. Rikhardsson, Martin Bennett, Jan Jaap Bouma and Stefan Schaltegger, 321–342. Dordrecht, The Netherlands: Springer. https://doi.org/10.1007/1-4020-3373-7_16.

Kokubu, Katsuhiko, and Hirotsugu Kitada. 2010. 'Conflicts and Solutions between Material Flow Cost Accounting and Conventional Management Thinking.' Paper presented at the 6th Asia-Pacific Interdisciplinary Perspectives on Accounting Research (APIRA) Conference, 12–13 July, University of Sydney. http://apira2010.econ.usyd.edu.au/conference_proceedings/APIRA-2010-150-Kokubu-Material-flow-cost-accounting.pdf.

Kolk, Ans, David Levy and Jonatan Pinkse. 2008. 'Corporate Responses in an Emerging Climate Regime: The Institutionalization and Commensuration of Carbon Disclosure.' *European Accounting Review* 17 (4): 719–745. https://doi.org/10.1080/09638180802489121.

Korpi, Eric, and Timo Ala-Risku. 2008. 'Life Cycle Costing: A Review of Published Case Studies.' *Managerial Auditing Journal* 23 (3): 240–261. https://doi.org/10.1108/02686900810857703.

Kücher, Alexander, Birgit Feldbauer-Durstmüller and Christine Duller. 2015. 'The Intellectual Foundations of Business Failure—A Co-Citation Analysis.' *Journal of International Business and Economics* 15 (2): 13–38. http://dx.doi.org/10.18374/JIBE-15-2.2.

Kuhn, Thomas S. 1996. *The Structure of Scientific Revolutions* (3rd edn). Chicago, US: University of Chicago Press.

Kundu, Sumit K., and Jerome A. Katz. 2003. 'Born-International SMEs: BI-Level Impacts of Resources and Intentions.' *Small Business Economics* 20 (1): 25–47. https://doi.org/10.1023/A:1020292320170.

Kvaløy, O. 2007. 'Asset Specificity and Vertical Integration.' *Scandinavian Journal of Economics* 109 (3): 551–572.

Labardin, Pierre, and Marc Nikitin. 2009. 'Accounting and the Words to Tell It: An Historical Perspective.' *Accounting, Business & Financial History* 19 (2): 149–166. https://doi.org/10.1080/09585200902969260.

Labuschagne, Carin, Alan C. Brent and Ron P. G. van Erck. 2005. 'Assessing the Sustainability Performances of Industries.' *Journal of Cleaner Production* 13 (4): 373–385. https://doi.org/10.1016/j.jclepro.2003.10.007.

Lane, Sarah J., and Martha Schary. 1991. 'Understanding the Business Failure Rate.' *Contemporary Economic Policy* 9 (4): 93–105. https://doi.org/10.1111/j.1465-7287.1991.tb00353.x.

Langfield-Smith, Kim. 2008. 'Strategic Management Accounting: How Far Have We Come in 25 Years?' *Accounting, Auditing & Accountability* 21 (2): 204–228. https://doi.org/10.1108/09513570810854400.

Laurans, Yann, Aleksander Rankovic, Raphaël Billé, Romain Pirard and Laurent Mermet. 2013. 'Use of Ecosystem Services Economic Valuation for Decision Making: Questioning a Literature Blindspot.' *Journal of Environmental Management* 119: 208–219. https://doi.org/10.1016/j.jenvman.2013.01.008.

Laurinkevičiūtė, Asta, Lorete Kinderytė and Žaneta Stasiškienė. 2008. 'Corporate Decision-Making in Furniture Industry: Weight of EMA and a Sustainability Balanced Scorecard.' *Environmental Research, Engineering and Management* 1 (43): 69–79. http://www.academia.edu/1023370/Corporate_Decision-Making_in_Furniture_Industry_Weight_of_EMA_and_a_Sustainability_Balanced_Scorecard.

Lee, K.-H. 2012. 'Carbon Accounting for Supply Chain Management in the Automobile Industry.' *Journal of Cleaner Production* 36: 83–93. https://doi.org/10.1016/j.jclepro.2012.02.023.

Lee, Ki-Hoon. 2011. 'Motivations, Barriers, and Incentives for Adopting Environmental Management (Cost) Accounting and Related Guidelines: A Study of the Republic of Korea.' *Corporate Social Responsibility and Environmental Management* 18 (1): 39–49. https://doi.org/10.1002/csr.239.

Lehnert, Christopher. 2011. 'Space Debris Removal for a Sustainable Space Environment.' *ESPI Perspectives* 52: Vienna, Austria: European Space Policy Institute, ESPI. https://espi.or.at/publications/voices-from-the-space-community/publications-of-the-former-espi-perspective-series/send/10-publications-of-the-former-espi-perspective-series/203-space-debris-removal-for-a-sustainable-space-environment.

Lehni, M. and World Business Council for Sustainable Development (WBCSD). 2000a. *Eco-efficiency: Creating More Value with Less Impact*. Geneva, Switzerland: WBCSD.

Li, Wenli. 2013. 'The Economics of Student Loan Borrowing and Repayment.' *Business Review* 2013 (3): 1–10. http://citeseerx.ist.psu.edu/viewdoc/download?doi=10.1.1.399.635&rep=rep1&type=pdf.

Lindholm, Anni, and Petri Suomala. 2007. 'Learning by Costing: Sharpening Cost Image through Life Cycle Costing?' *International Journal of Productivity and Performance Management* 56 (8): 651–672. https://doi.org/10.1108/17410400710832985.

Liou, J.-C. 2011. 'Engineering and Technology Challenges for Active Debris Removal.' *EUCASS Proceedings Series* 4: 735–748. Saint Petersburg, Russia. http://dx.doi.org/10.1051/eucass/201304735.

Llena, Fernando, José M. Moneva and Blanca Hernandez. 2007. 'Environmental Disclosures and Compulsory Accounting Standards: The Case of Spanish Annual Reports.' *Business Strategy and the Environment* 16 (1): 50–63. https://doi.org/10.1002/bse.466.

Lodhia, Sumit K. 2003. 'Accountants' Responses to the Environmental Agenda in a Developing Nation: An Initial and Exploratory Study on Fiji.' *Critical Perspectives on Accounting* 14 (7): 715–737. https://doi.org/10.1016/S1045-2354(02)00190-9.

Lundberg, Erik. 1969. Award ceremony speech for the Sveriges Riksbank Prize in Economic Sciences in Memory of Alfred Nobel, 1969, given to Ragnar Frisch and Jan Tinbergen. https://www.nobelprize.org/prizes/economic-sciences/1969/ceremony-speech/.

Lungu, Camelia Iuliana, Chiraţa Caraiani, Cornelia Dascalu and Gina Raluca Guşe. 2009. 'Social and Environmental Determinants of Risk and Uncertainties Reporting.' *Issues in Social and Environmental Accounting* 3 (2): 100–116. 10.22164/isea.v3i2.39.

Lussier, Robert N. 1995. 'A Nonfinancial Business Success versus Failure Prediction Model for Young Firms.' *Journal of Small Business Management* 33 (1): 8–20. https://www.questia.com/library/journal/1G1-16787401/a-nonfinancial-business-success-versus-failure-prediction.

Lyytinen, Kalle, and Jan Damsgaard. 2001. 'What's Wrong with the Diffusion of Innovation Theory?' In *Diffusing Software Product and Process Innovations*, edited by Mark A. Ardis and Barbara L. Marcolin, IFIP—The International Federation for Information Processing book series 59: 173–190. Conference proceedings, TDIT 2001, April 7–10, Banff, Canada. Boston, MA: Springer. https://doi.org/10.1007/978-0-387-35404-0_11.

Machado, André, Fabiano Guasti Lima and Antonio Carlos da Silva Filho. C. 2011. 'Carbon Credit Storage: A Study of How to Measure and Account Posting.' *Review of Business Research* 11 (1): 126–133. https://www.researchgate.net/publication/279752559_CARBON_CREDIT_STORAGE_A_STUDY_OF_HOW_TO_MEASURE_AND_ACCOUNT_POSTING.

Mahadevia, Darshini, and Jeanne Wolfe. 2008. *Solid Waste Management in Indian Cities: Status and Emerging Practices.* New Delhi, India: Concept Publishing Company.

Mahmood, Shakeel Ahmed Ibne, and Amal Krishna Halder. 2011. 'The Socioeconomic Impact of Arsenic Poisoning in Bangladesh.' *Journal of Toxicology and Environmental Health Sciences* 3 (3): 65–73. https://academicjournals.org/journal/JTEHS/article-abstract/1AEF32A1062.

Malarvizhi, P., and Sangeeta Yadav. 2008. 'Corporate Environmental Disclosures on the Internet: An Empirical Analysis of Indian Companies.' *Issues in Social and Environmental Accounting* 2 (2): 211–232. https://doi.org/10.22164/isea.v2i2.33.

Mann, Shamsher Singh, and Deepika Thadani. 2010. 'ECOTEL Version 2.0—Reaching Out.' 12 April 2010, hvs.com. http://www.hvs.com/article/4492/ecotelversion20reachingout.

Mattessich, Richard. 1971. 'The Market Value Method According to Sterling: A Review Article.' *Abacus* 7 (2): 176–193. https://doi.org/10.1111/j.1467-6281.1971.tb00405.x.

Mayer, Audrey L. 2008. 'Strengths and Weaknesses of Common Sustainability Indices for Multidimensional Systems.' *Environment International* 34 (2): 277–291. https://doi.org/10.1016/j.envint.2007.09.004.

McCully, Patrick. 2001. *Silenced Rivers: The Ecology and Politics of Large Dams.* New York: Zed Books.

McGrath, Rita Dorothea Gunther. 1999. 'Falling Forward: Real Options Reasoning and Entrepreneurial Failure.' *Academy of Management Review* 24 (1): 13–30. https://doi.org/10.5465/amr.1999.1580438.

McLaughlin, Andrew. 1993. *Regarding Nature: Industrialism and Deep Ecology.* Albany, NY: State University of New York Press.

Meij, Ruud. 2003. *Status Report on the Health Issues Associated with Pulverised Fuel Ash and Fly Dust.* Arnhem, The Netherlands: KEMA Power Generation & Sustainables.

Meinking, Mary. 2014. *Cash Crop to Cash Cow: The History of Tobacco and Smoking in America.* Broomall, US: Mason Crest.

Merino, Barbara D. 1993. 'An Analysis of the Development of Accounting Knowledge: A Pragmatic Approach.' *Accounting, Organizations and Society* 18 (2): 163–185. https://doi.org/10.1016/0361-3682(93)90032-2.

Metcalf, Louise, and Suzanne Benn. 2012. 'The Corporation is Ailing Social Technology: Creating a "Fit for Purpose" Design for Sustainability.' *Journal of Business Ethics* 111 (2): 195–210. https://doi.org/10.1007/s10551-012-1201-1.

Milgram, Stanley. 1963. 'Behavioral Study of Obedience.' *Journal of Abnormal and Social Psychology* 67 (4): 371–378. https://doi.org/10.1037/h0040525.

Millennium Ecosystem Assessment. 2005. *Ecosystems and Human Well-Being.* Washington DC, US: Island Press.

Ministry of the Environment (MOE) Japan. 2005. *Environmental Accounting Guidelines.* MOE Japan. https://www.env.go.jp/en/policy/ssee/eag05.pdf.

Moneva, José M., and Eduardo Ortas. 2010. 'Corporate Environmental and Financial Performance: A Multivariate Approach.' *Industrial Management & Data Systems* 110 (2): 193–210. https://doi.org/10.1108/02635571011020304.

Moosa, Imad A., and Lee Smith. 2004. 'Economic Development Indicators as Determinants of Medal Winning at the Sydney Olympics: An Extreme Bounds Analysis.' *Australian Economic Papers* 43 (3): 288–301. https://doi.org/10.1111/j.1467-8454.2004.00231.x.

Mori, K., and A. Christodoulou. 2012. 'Review of Sustainability Indices and Indicators: Towards a New City Sustainability Index (CSI)'. *Environmental Impact Assessment Review*, 32(1): 94–106. doi: 10.1016/j.eiar.2011.06.001

Nakajima, M. 2011. 'Environmental management Accounting for Cleaner Production: Systematization of Material Flow Cost Accounting (MFCA) into Corporate Management System.' *Kansai University Review of Business & Commerce* 13: 17–39. https://kuir.jm.kansai-u.ac.jp/dspace/bitstream/10112/4731/1/KU-1100-201103-02.pdf.

Nakajima, Michiyasu. 2009. 'Evolution of Material Flow Cost Accounting (MFCA): Characteristics on Development of MFCA Companies and Significance and Relevance of MFCA.' *Kansai University Review of Business and Commerce* 11: 27–46. https://kuir.jm.kansai-u.ac.jp/dspace/bitstream/10112/903/1/KU-1100-20090300-03.pdf.

National Research Council. 2010. *Hidden Costs of Energy: Unpriced Consequences of Energy Production and Use*. Washington, DC: The National Academies Press. https://doi.org/10.17226/12794.

National Research Council. 2011. *Limiting Future Collision Risk to Spacecraft: An Assessment of NASA's Meteoroid and Orbital Debris Programs*. Washington, DC: The National Academies Press. https://doi.org/10.17226/13244.

Negash, Minga. 2012. 'IFRS and Environmental Accounting.' *Management Research Review* 35 (7): 577–601. https://doi.org/10.1108/01409171211238811.

New South Wales (NSW) government. 2011. *Energy Saver—Technology Report: Industrial Refrigeration and Chilled Glycol and Water Applications*. Sydney, NSW, Australia: Office of Environment and Heritage, Department of Premier and Cabinet, State of NSW. https://www.environment.nsw.gov.au/resources/business/110302-industrial-refrigeration-tech-rpt.pdf

Nigam, B. M. Lall. (1986). 'Bahi-Khata: The Pre-Pacioli Indian Double-entry System of Bookkeeping.' *Abacus* 22 (2): 148–161. https://doi.org/10.1111/j.1467-6281.1986.tb00132.x.

Nikolaou, Ioannis E., and Konstantinos I. Evangelinos. 2010. 'Classifying Current Social Responsibility Accounting Methods for Assisting a Dialogue between Business and Society.' *Social Responsibility Journal* 6 (4): 562–579. https://doi.org/10.1108/17471111011083446.

Nixon, B., and J. Burns. 2012b. 'The Paradox of Strategic Management Accounting.' *Management Accounting Research* 23 (4): 229–244. https://doi.org/10.1016/j.mar.2012.09.004.

Nixon, Bill, and John Burns. 2012a. 'Editorial: Strategic Management Accounting.' *Management Accounting Research* 23 (4): 225–228. https://doi.org/10.1016/j.mar.2012.09.005.

Norris, Gregory A. 2001. 'Integrating Life Cycle Cost Analysis and LCA.' *International Journal of LCA* 6 (2): 118–120. https://doi.org/10.1007/BF02977849.

Nyquist, Siv. 2003. 'Environmental Information in Annual Reports: A Survey of Swedish Accountants.' *Managerial Auditing Journal* 18 (8): 682–691. https://doi.org/10.1108/02686900310495935.

Odum, Eugene P., and Gary W. Barrett. 2005. *Fundamentals of Ecology* (5th edn). Belmont, CA: Thomson Brooks/Cole.

Ogilvy, Sue. 2015. 'Developing the Ecological Balance Sheet for Agricultural Sustainability.' *Sustainability Accounting, Management and Policy Journal* 6 (2): 110–137.

Ojea, Elena, Julia Martin-Ortega and Aline Chiabai. 2012. 'Defining and Classifying Ecosystem Services for Economic Valuation: The Case of Forest Water Services.' *Environmental Science & Policy* 19–20: 1–15. https://doi.org/10.1016/j.envsci.2012.02.002.

Onishi, Yasushi, Katsuhiko Kokubu and Michiyasu Nakajima. 2009. 'Implementing Material Flow Cost Accounting in a Pharmaceutical Company.' In *Environmental Management Accounting for Cleaner Production*, edited by

Stefan Schaltegger, Martin Bennett, Roger L. Burritt and Christine Jasch, *Eco-Efficiency in Industry and Science* 24, 395–409. Dordrecht: Springer. https://doi.org/10.1007/978-1-4020-8913-8_22.

Orlikowski, Wanda J. 1992. 'The Duality of Technology: Rethinking the Concept of Technology in Organizations.' *Organization Science* 3 (3): 398–427. https://doi.org/10.1287/orsc.3.3.398.

Orlitzky, Marc, and Glen Whelan. 2007. 'On the Effectiveness of Social and Environmental Accounting.' *Issues in Social and Environmental Accounting* 1 (2): 311–333. https://doi.org/10.22164/isea.v1i2.20.

Ortiz, Ramon Arigoni, Alexander Golub, Oleg Lugovoy, Anil Markandya and James Wang. 2011. 'DICER: A Tool for Analyzing Climate Policies.' *Energy Economics* 33 (supplement 1): S41–S49. https://doi.org/10.1016/j.eneco.2011.07.025.

Pandey, V. C., Jay Shankar Singh, Rana P. Singh, Nandita Singh and M. Yunus. 2011. 'Arsenic Hazards in Coal Fly Ash and its Fate in Indian Scenario.' *Resources, Conservation and Recycling* 55 (9–10): 819–835. https://doi.org/10.1016/j.resconrec.2011.04.005.

Pandey, Vimal Chandra, and Nandita Singh. 2010. 'Impact of Fly Ash Incorporation in Soil Systems.' *Agriculture, Ecosystems and Environment* 136 (1–2): 16–27. https://doi.org/10.1016/j.agee.2009.11.013.

Papaspyropoulos, Konstantinos G., Vaios Blioumis, Athanassios S. Christodoulou, Periklis K. Birtsas and Kyriakos E. Skordas. 2012. 'Challenges in Implementing Environmental Management Accounting Tools: The Case of a Nonprofit Forestry Organization.' *Journal of Cleaner Production* 29–30: 132–143. https://doi.org/10.1016/j.jclepro.2012.02.004.

Parker, L. D. 2011. 'Twenty-One Years of Social and Environmental Accountability Research: A Coming of Age.' *Accounting Forum* 35 (1): 1–10. https://doi.org/10.1016/j.accfor.2010.11.001.

Parker, L. D., James Guthrie, J., & and Simon Linacre, S. (2011). Editorial: 'The relationship Relationship between academic Academic accounting Accounting research Research and professional Professional practice Practice.' *Accounting, Auditing & Accountability Journal*, 24 (1): 5–14. https://doi.org/10.1108/09513571111098036.

Parker, Lee D. 2005. 'Social and Environmental Accountability Research: A View from the Commentary Box.' *Accounting, Auditing & Accountability Journal* 18 (6): 842–860. https://doi.org/10.1108/09513570510627739.

Paton, William A. 1922. 'Valuation of Inventories.' *Journal of Accountancy* 34: 432–450.

Pedersen, Ole Gravgård, and Mark de Haan. 2006. 'The System of Environmental and Economic Accounts—2003 and the Economic Relevance of Physical Flow Accounting.' *Journal of Industrial Ecology* 10 (1–2): 19–42. https://doi.org/10.1162/108819806775545466.

Petcharat, Neungruthai Nickie, and Joseph M. Mula. 2012. 'Towards a Conceptual Design for Environmental and Social Cost Identification and Measurement

System.' *Journal of Financial Reporting & Accounting* 10 (1): 34–54. https://doi.org/10.1108/19852511211237435.

Peters-Stanley, Molly, Katherine Hamilton, Thomas Marcello and Milo Sjardin. 2011. *Back to the Future: State of the Voluntary Carbon Markets 2011.* New York and Washington, DC: Ecosystem Marketplace & Bloomberg New Energy Finance. https://www.forest-trends.org/publications/back-to-the-future

Pfeffer, Jeffrey, and Gerald R. Salancik. 1978. *The External Control of Organizations: A Resource Dependence Perspective.* New York: Harper & Row.

Pirard, Romain. 2012. 'Market-based Instruments for Biodiversity and Ecosystem Services: A Lexicon.' *Environmental Science & Policy* 19–20: 59–68. https://doi.org/10.1016/j.envsci.2012.02.001.

Pongrácz, Eva, and Veikko J. Pohjola. 2004. 'Re-defining Waste, the Concept of Ownership and the Role of Waste Management.' *Resources, Conservation & Recycling* 40 (2): 141–153. https://doi.org/10.1016/S0921-3449(03)00057-0.

Prasad, Bably, and Kajal Kumar Mondal. 2008. 'The Impact of Filling an Abandoned Open Cast Mine with Fly Ash on Ground Water Quality: A Case Study.' *Mine Water and the Environment* 27 (1): 40–45. https://doi.org/10.1007/s10230-007-0021-5.

Prasad, N. K. 1977. *Principles and Practice of Cost Accounting.* Calcutta, India: Book Syndicate. https://archive.org/details/in.ernet.dli.2015.460233/page/n2.

Proctor, Robert N. 2000. *The Nazi War on Cancer.* New Jersey, US: Princeton University Press.

Pulver, Simone. 2007. 'Introduction: Developing-Country Firms as Agents of Environmental Sustainability?' *Studies in Comparative International Development* 42 (3–4): 191–207. https://doi.org/10.1007/s12116-007-9011-7.

Quintana, M. J. M., Gallego, A. G., and Pascual, M. E. V. 2012. 'Análisis del fracaso empresarial por sectores: factores diferenciadores' [Cross-industry Analysis of Business Failure: Differential Factors]. *Pecvnia: Revista de la Facultad de Ciencias Económicas y Empresariales, Universidad de León:* 53–83.

Rabl, Ari, and Joseph V. Spadaro. 2000. 'Health Costs of Automobile Pollution.' *Revue Française d'Allergologie et d'Immunologie Clinique* 40 (1): 55–59. https://doi.org/10.1016/S0335-7457(00)80021-0.

Radwan, Hatem R. I., Eleri Jones and Dino Minoli. 2010. 'Managing Solid Waste in Small Hotels.' *Journal of Sustainable Tourism* 18 (2): 175–190. https://doi.org/10.1080/09669580903373946.

Ramachandra, T. V., and Shwetmala. 2009. 'Emissions from India's Transport Sector: Statewise Synthesis.' *Atmospheric Environment* 43 (34): 5510–5517. https://doi.org/10.1016/j.atmosenv.2009.07.015.

Rametsteiner, Ewald, Helga Pülzl, Johanna Alkan-Olsson and Pia Frederiksen. 2011. 'Sustainability Indicator Development—Science or Political Negotiation?' *Ecological Indicators* 11 (1): 61–70. https://doi.org/10.1016/j.ecolind.2009.06.009.

Ramos, Tomás B. 2009. 'Development of Regional Sustainability Indicators and the Role of Academia in this Process: The Portuguese Practice.' *Journal of Cleaner Production* 17 (12): 1101–1115. https://doi.org/10.1016/j.jclepro.2009.02.024.

Rao, Purba, Alok Kumar Singh, Olivia la O'Castillo, Ponciano S. Intal, Jr., and Ather Sajid. 2009. 'A Metric for Corporate Environmental Indicators.... for Small and Medium Enterprises in the Philippines.' *Business Strategy and the Environment* 18 (1): 14–31. https://doi.org/10.1002/bse.555.

Ratnatunga, Janek T. D., and Kashi R. Balachandran. 2009. 'Carbon Business Accounting: The Impact of Global Warming on the Cost and Management Accounting Profession.' *Journal of Accounting, Auditing & Finance* 24 (2): 333–355. https://doi.org/10.1177%2F0148558X0902400208.

Ratnatunga, J. T. D., Stewart Jones and K. R. Balachandran. 2011. 'The Valuation and Reporting of Organizational Capability in Carbon Emissions Management.' *Accounting Horizons* 25 (1): 127–147. https://doi.org/10.2308/acch.2011.25.1.127.

Rebitzer, G., Tomas Ekvall, Rolf Frischknecht, D. Hunkeler, Gregory A. Norris, Tomas Rydberg, W.-P. Schmidt, Sangwon Suh, Bo Pedersen Weidema and D. W. Pennington. 2004. 'Life Cycle Assessment Part 1: Framework, Goal and Scope Definition, Inventory Analysis, and Applications.' *Environment International* 30 (5): 701–720.

Reierson, A and M. Carr. 2018. 'Carbon Options Signal 20% Gain as Europe Nears Record Price.' Retrieved from https://www.bloomberg.com/news/articles/2018-09-10/carbon-traders-bet-on-rally-going-much-further-options-show

Reiner, Rob. (producer, director). 1992. *A Few Good Men*. Los Angeles, California, US: Castle Rock Entertainment.

Renfrew, C. 1977. 'The Later Obsidian of Deh Luran—The Evidence of Chagha Sefid.' In *Studies in the Archaeological History of the Deh Luran Plain: The Excavation of Chagha Sefid* by Frank Hole, with contributions by M. J. Kirkby and Colin Renfrew, *Memoirs of the Museum of Anthropology* 9: 289–311. Ann Arbor: Museum of Anthropology, University of Michigan.

Renfrew, Colin. 1969. 'Trade and Culture Process in European Prehistory.' *Current Anthropology* 10 (2/3): 151–169. http://doi.org/10.1086/201066.

Repetto, Robert, William Magrath, Michael Wells, Christine Beer and Fabrizio Rossini. 1989. *Wasting Assets: Natural Resources in the National Income Accounts.* Washington, DC: World Resources Institute.

Richard, Jacques. 2017. 'The Need to Reform the Dangerous IFRS System of Accounting.' *Accounting, Economics, and Law: A Convivium* 7 (2): 93–103. https://doi.org/10.1515/ael-2017-0017.

Robkob, Phaiboon, and Phapurke Ussahawanitchakit. 2009. 'Antecedents and Consequences of Voluntary Disclosure of Environmental Accounting: An Empirical Study of Foods and Beverage Firms in Thailand.' *Review of Business Research* 9 (3): 1–30.

Rogers, M. E. 1983. *Diffusion of Innovations* (3rd edn). New York: Free Press.

Rosendahl, Knut Einar, and Jon Strand. 2011. 'Carbon Leakage from the Clean Development Mechanism.' *The Energy Journal* 32 (4): 27–50. https://doi.org/10.5547/ISSN0195-6574-EJ-Vol32-No4-3.

Rout, H.S. 2010. 'Green Accounting: Issues and Challenges'. *The IUP Journal of Managerial Economics*, 8(3), 46–60.

Rowe, Christopher L., William A. Hopkins and Justin D. Congdon. 2002. 'Ecotoxicological Implications of Aquatic Disposal of Coal Combustion Residues in the United States: A Review.' *Environmental Monitoring and Assessment* 80 (3): 207–276. http://doi.org/10.1023/A:1021127120575.

Rout, Himanshu Sekhar. 2010. 'Green Accounting: Issues and Challenges.' *The IUP Journal of Managerial Economics* 8 (3): 46–60. https://ssrn.com/abstract=1681949.

Rubinstein, Ariel. 2006. 'A Sceptic's Comment on the Study of Economics.' *The Economic Journal* 116 (510): C1–C9. https://doi.org/10.1111/j.1468-0297.2006.01071.x.

Ruhl, Laura, Avner Vengosh, Gary S. Dwyer, Heileen Hsu-Kim, Grace Schwartz, Autumn Romanski and S. Daniel Smith. 2012. 'The Impact of Coal Combustion Residue Effluent on Water Resources: A North Carolina Example.' *Environmental Science & Technology* 46 (21), 12226–12233. https://doi.org/10.1021/es303263x.

Ruzzier, C. 2009. 'Asset Specificity and Vertical Integration: Williamson's Hypothesis Reconsidered.' HBS Working Paper 09-119, Harvard Business School.

Sagoff, Mark. 2011. 'The Quantification and Valuation of Ecosystem Services.' *Ecological Economics* 70 (3): 497–502. https://doi.org/10.1016/j.ecolecon.2010.10.006.

Saini, S., Rao, P., Patil, Y. 2012. 'City based Analysis of MSW to Energy Generation in India Calculation of State-wise Potential and Tariff Comparison with EU.' *Procedia: Social and Behavioral Sciences*, 37, 407–416. doi:10.1016/j.sbspro.2012.03.306

Saito, Shizuki, and Yoshitaka Fukui. 2016. 'Whither the Concept of Income?' *Accounting, Economics, and Law: A Convivium.* http://doi.org/10.1515/ael-2016-0013.

Sarkar, A. N. 2010. 'Emissions Trading and Carbon Credit Accounting For Sustainable Energy Development with Focus on India.' *Globsyn Management Journal* 4 (1–2): 35–62.

Schaltegger, S., and Maria Csutora. 2012. 'Carbon Accounting for Sustainability and Management: Status Quo and Challenges.' *Journal of Cleaner Production* 36: 1–16. https://doi.org/10.1016/j.jclepro.2012.06.024.

Schaltegger, S., Tobias Viere and Dimitar Zvezdov. 2012. 'Tapping Environmental Accounting Potentials of Beer Brewing: Information Needs for Successful Cleaner Production.' *Journal of Cleaner Production* 29–30: 1–10. https://doi.org/10.1016/j.jclepro.2012.02.011.

Schaltegger, Stefan. 1997. 'Information Costs, Quality of Information and Stakeholder Involvement—The Necessity of International Standards of Ecological Accounting.' *Eco-Management and Auditing* 4 (3): 87–97. https://doi.org/10.1002/(SICI)1099-0925(199711)4:3%3C87::AID-EMA70%3E3.0.CO;2-Z.

Scheer, August-Wilhelm, and Markus Nüttgens. 2000. 'ARIS Architecture and Reference Models for Business Process Management.' In *Business Process Management* (3rd edn), edited by Wil van der Aalst, Jörg Desel and Andreas Oberweis, 376–389. Berlin, Germany: Springer.

Schmied, M., and W. Knörr. 2012. *Calculating GHG Emissions for Freight Forwarding and Logistics Services in Accordance with EN 16258*. Germany, European Association for Forwarding, Transport, Logistics and Customs Services (CLECAT).

Schultz, Karl, and Peter Williamson. 2005. 'Gaining Competitive Advantage in a Carbon-Constrained World: Strategies for European Business.' *European Management Journal* 23 (4): 383–391. https://doi.org/10.1016/j.emj.2005.06.010.

Schultz, E.T., R.J. Johnston, K. Segerson and E.Y. Besedin. 2012. 'Integrating Ecology and Economics for Restoration, using Ecological Indicators in Valuation of Ecosystem Services'. *Restoration Ecology*, 20(3): 304–310. doi:10.1111/j.1526-100X.2011.00854.x

Schwitzgebel, Eric. 2002. 'Why Did We Think We Dreamed in Black and White?' *Studies in History and Philosophy of Science Part A* 33 (4): 649–660. https://doi.org/10.1016/S0039-3681(02)00033-X

Seiler-Hausmann, Jan-Dirk, Christa Liedtke and Ernst Ulrich von Weizsäcker. 2004. *Eco-efficiency and Beyond: Towards the Sustainable Enterprise*. Sheffield, UK: Greenleaf Publishing.

Sepúlveda, A., Schluep, M., Renaud, F.G., Streicher, M., Kuehr, R., Hagelüken, C., et al. 2010. 'A Review of the Environmental Fate and Effects of Hazardous Substances Released from Electrical and Electronic Equipments during Recycling: Examples from China and India.' *Environmental Impact Assessment Review*, 30, 28–41. doi:10.1016/j.eiar.2009.04.001

Shamsad, A., M.H. Fulekar and B. Pathak. 2012. 'Impact of Coal Based Thermal Power Plant on Environment and its Mitigation Measure.' *International Research Journal of Environment Sciences*, 1(4): 60–64.

Shank, John K. 2006. 'Strategic Cost Management: Upsizing, Downsizing, and Right (?) Sizing.' In *Contemporary Issues in Management Accounting*, edited by Alnoor Bhimani, 355–379. Oxford, UK: Oxford University Press. https://doi.org/10.1093/acprof:oso/9780199283361.003.0016.

Sharholy, M., Ahmad, K., Mahmood, G., Trivedi, R.C. 2008. 'Municipal Solid Waste Management in Indian Cities—A Review.' *Waste Management*, 28, 459–467. doi:10.1016/j.wasman.2007.02.008

Sharma, R. N., and Shashi R. Singh. 2009. 'Displacement in Singrauli Region: Entitlements and Rehabilitation.' *Economic & Political Weekly* 44 (51): 62–69.

Shen, Li-Yin, J. Jorge Ochoa, Mona N. Shah and Xiaoling Zhang. 2011. 'The Application of Urban Sustainability Indicators—A Comparison between Various Practices.' *Habitat International* 35 (1): 17–29. https://doi.org/10.1016/j.habitatint.2010.03.006.

Shevchuk, Volodymyr. 2013. 'Assets as Accounting, Control and Analysis Objects: Ecology & Economic Identification and Interpretation.' *Accounting and Finance* 1: 66–73.

Sidhu, Balraj. 2011. 'The Niyamgiri Hills Bauxite Project: Balancing Resource Extraction and Environment Protection.' *Environmental Policy and Law* 41 (3): 166–171.

Simon, Herbert A. 1956. 'Rational Choice and the Structure of the Environment.' *Psychological Review* 63 (2): 129–138. https://doi.org/10.1037/h0042769.

Singh, Gurdip, and Malar Joshi. 2009. 'Environment Management and Disclosure Practices of Indian Companies.' *International Journal of Business Research* 9 (2): 116–128.

Singh, Gurdeep, S. K. Gupta, Ritesh Kumar and M. Sunderarajan. 2007. 'Mathematical Modeling of Leachates from Ash Ponds of Thermal Power Plants.' *Environmental Monitoring & Assessment* 130 (1–3): 173–185. https://doi.org/10.1007/s10661-006-9387-2.

Singh, Jay Shankar, and Vimal Chandra Pandey.). 'Fly Ash Application in Nutrient Poor Agriculture Soils: Impact on Methanotrophs Population Dynamics and Paddy Yields.' *Ecotoxicology and Environmental Safety* 89: 43–51. https://doi.org/10.1016/j.ecoenv.2012.11.011.

Sjöberg, Lennart, and Britt-Marie Drottz-Sjöberg. 2009. 'Public Risk Perception of Nuclear Waste.' *International Journal of Risk Assessment and Management* 11 (3–4): 264–296. https://doi.org/10.1504/IJRAM.2009.023156.

Smith, Allan H., Elena O. Lingas and Mahfuzar Rahman. 2000. 'Contamination of Drinking-Water by Arsenic in Bangladesh: A Public Health Emergency.' *Bulletin of the World Health Organization* 78 (9): 1093–1103.

Sorvari, Jaana, and Jyri Seppälä. 2010. 'A Decision Support Tool to Prioritize Risk Management Options for Contaminated Sites.' *Science of the Total Environment* 408 (8): 1786–1799. https://doi.org/10.1016/j.scitotenv.2009.12.026.

Spash, C.L. 2008. 'How Much is that Ecosystem in the Window? The One with the Bio-diverse Trail.' *Environmental Values*, 17(2): 259–284.

Srivastava, Samir K. 2007. 'Green Supply-Chain Management: A State-of-the-Art Literature Review.' *International Journal of Management Reviews (IJMR)* 9 (1): 53–80. https://doi.org/10.1111/j.1468-2370.2007.00202.x.

Staniskis, Jurgis Kazimieras, and Zaneta Stasiskiene. 2006. 'Environmental Management Accounting in Lithuania: Exploratory Study of Current Practices, Opportunities and Strategic Intents.' *Journal of Cleaner Production* 14 (14): 1252–1261. https://doi.org/10.1016/j.jclepro.2005.08.009.

Stasiškienė, Žaneta, and Rasa Juškaitė. 2007. 'Optimization of the Cardboard Manufacturing Process in Accordance with Environmental and Economic

Factors.' *Environmental Research, Engineering and Management* 2 (40): 70–79. https://www.academia.edu/1023377/Optimization_of_the_Cardboard_Manufacturing_Process_in_Accordance_with_Environmental_and_Economic_Factors.

Steen, Bengt. 2005. 'Environmental Costs and Benefits in Life Cycle Costing.' *Management of Environmental Quality: An International Journal* 16 (2). 107–118. https://doi.org/10.1108/14777830510583128.

Stegall, Nathan. 2006. 'Designing for Sustainability: A Philosophy for Ecologically Intentional Design.' *Design Issues* 22 (2): 56–63. https://doi.org/10.1162/desi.2006.22.2.56.

Sterner, Eva. 2001. 'Life-Cycle Costing and its Use in the Swedish Building Sector.' *Building Research & Information* 28 (5–6), 387–393. https://doi.org/10.1080/096132100418537.

Stinchcombe, Arthur. 1965. 'Organization-Creating Organizations.' *Trans-action* 2 (2): 34–35. https://doi.org/10.1007/BF03180801.

Sullivan, Sian, and Mike Hannis. 2017. '"Mathematics Maybe, But Not Money": On Balance Sheets, Numbers and Nature in Ecological Accounting.' *Accounting, Auditing & Accountability Journal* 30 (7): 1459–1480. https://doi.org/10.1108/AAAJ-06-2017-2963.

Sushil, Snigdha, and Vidya S. Batra. 2006. 'Analysis of Fly Ash Heavy Metal Content and Disposal in Three Thermal Power Plants in India.' *Fuel* 85 (17–18): 2676–2679. https://doi.org/10.1016/j.fuel.2006.04.031.

Suzuki, Tomo. 2003. 'The Epistemology of Macroeconomic Reality: The Keynesian Revolution from an Accounting Point of View.' *Accounting, Organizations and Society* 28 (5): 471–517. http://doi.org/10.1016/S0361-3682(01)00061-7.

TERI. 2006. 'Policy, Institutional and Legal Barriers to Economic Utilisation of Fly Ash.' TERI. http://www.teriin.org/upfiles/projects/ (last accessed 15 January 2017).

Thornhill, Stewart, and Raphael Amit. 2003. 'Learning about Failure: Bankruptcy, Firm Age, and the Resource-Based View.' *Organization Science* 14 (5): 497–509. https://doi.org/10.1287/orsc.14.5.497.16761.

Tol, Richard S. J. 2005. 'The Marginal Damage Costs of Carbon Dioxide Emissions: An Assessment of Uncertainties.' *Energy Policy* 33 (16): 2064–2074. https://doi.org/10.1016/j.enpol.2004.04.002.

Tukker, Arnold. 2000. 'Life Cycle Assessment as a Tool in Environmental Impact Assessment.' *Environmental Impact Assessment Review* 20 (4): 435–456. https://doi.org/10.1016/S0195-9255(99)00045-1.

Turcu, Catalina. 2013. 'Re-thinking Sustainability Indicators: Local Perspectives of Urban Sustainability.' *Journal of Environmental Planning and Management* 56 (5): 695–719. https://doi.org/10.1080/09640568.2012.698984.

US Congress, Office of Technology Assessment (OTA). 1990. *Orbiting Debris: A Space Environmental Problem*. Background paper OTA-BP-ISC-72. Washington, DC: US Government Printing. https://ota.fas.org/reports/9033.pdf.

Ulrich, Peter. 2010. 'Civilizing the Market Economy: The Approach of Integrative Economic Ethics to Sustainable Development.' *Economics, Management, and Financial Markets (official Journal of the Contemporary Science Association)*, 5 (1). 99–112. Available at: https://www.alexandria.unisg.ch/63739/

Unerman, Jeffrey, and Brendan O'Dwyer. 2007. 'The Business Case for Regulation of Corporate Social Responsibility and Accountability.' *Accounting Forum* 31 (4): 332–353. https://doi.org/10.1016/j.accfor.2007.08.002.

UNFCCC. 2009. 'Boiler Fuel Conversion from RFO to Biomass Based Briquettes at Fresenius Kabi India Private Limited, Ranjangaon (M.S.), India'; UNFCCC Project 1497. http://cdm.unfccc.int/Projects/DB/DNV-CUK1199787528.27/view.

United Nations Division for Sustainable Development (UNDSD). 2001. *Environmental Management Accounting Procedures and Principles*. New York: UN.

United Nations Environment Programme (UNEP). 2012. 'Trends in Hotel Certifications and Rating Programs: Guidelines for the Caribbean.' Caribbean Environment Network Project (CEN). http://www.cep.unep.org/issues/hotel_cert.PDF.

United Nations Environmental Protection Agency (US EPA). 1995. *An Introduction to Environmental Accounting as a Business Management Tool: Key Concepts and Terms*. Washington DC, US: Office of Pollution Prevention and Toxics, US EPA.

United Nations Framework Convention of Climate Change (UNFCCC). 2008. *Kyoto Protocol Reference Manual: On Accounting of Emissions and Assigned Amount*. Bonn, Germany: Climate Change Secretariat, UNFCCC. https://unfccc.int/resource/docs/publications/08_unfccc_kp_ref_manual.pdf.

van den Bergh, Jeroen C. 2001. 'Ecological Economics: Themes, Approaches, and Differences with Environmental Economics.' *Regional Environmental Change* 2 (1): 13–23. https://doi.org/10.1007/s101130000020.

van der Meer-Kooistra, Jeltje, and Ed G. J. Vosselman. 2012. 'Research Paradigms, Theoretical Pluralism and the Practical Relevance of Management Accounting Knowledge.' *Qualitative Research in Accounting & Management* 9 (3): 245–264. https://doi.org/10.1108/11766091211257452.

Vannella, Giovanni. 2004. 'Problems Related to the Development of Integrated Systems for Economic and Environmental Accounting: A Preliminary Analysis of the Economic-Environmental Impact of Human Activities on the Marine Pollution in the Basilicata Region.' *Statistica* 64 (3): 587–598. https://doi.org/10.6092/issn.1973-2201/60

Velasquez, Mark, and Patrick T. Hester. 2013. 'An Analysis of Multi-Criteria Decision Making Methods.' *International Journal of Operations Research* 10 (2): 56–66. https://www.orstw.org.tw/ijor/vol10no2/ijor_vol10_no2_p56_p66.pdf

Verrecchia, Robert E. 2013. 'Accounting Alchemy.' *Accounting Horizons* 27 (3): 603–617. https://doi.org/10.2308/acch-50488.

Viere, Tobias, Stefan Schaltegger and Jan von Enden. 2007. 'Supply Chain Information in Environmental Management Accounting—The Case of a Vietnamese Coffee Exporter.' *Issues in Social and Environmental Accounting* 1 (2): 296–310. https://doi.org/10.22164/isea.v1i2.19.

Warrier, A. G. Krishna. 1953. *Maha Upanishad*, verses VI.71–72. Madras: Theosophical Society. https://www.advaita.it/library/mahaupanishad.htm.

Watson, Peter. 2009. *Ideas: A History of Thought and Invention, from Fire to Freud.* Harper Collins.

WBCSD. 2000b. *Measuring Eco-efficiency: A Guide to Reporting Company Performance.* Geneva, Switzerland: WBCSD. http://www.gdrc.org/sustbiz/measuring.pdf.

WBCSD and World Resources Initiative (WRI). 2004. *The Greenhouse Gas Protocol: A Corporate Accounting and Reporting Standard*, revised edition. Geneva, Switzerland: WBCSD; Washington, DC, USA: WRI. https://www.wri.org/publication/greenhouse-gas-protocol.

Whyte, G. 1994. 'The Role of Asset Specificity in the Vertical Integration Decision.' *Journal of Economic Behavior & Organization*, 23 (3): 287–302. doi: 10.1016/0167-2681(94)90003-5.

Wilburn, Kathleen, and Ralph Wilburn. 2013. 'Using Global Reporting Initiative Indicators for CSR Programs.' *Journal of Global Responsibility*, 4 (1): 62–75. https://doi.org/10.1108/20412561311324078.

Williams, M. 2008. 'Safeguarding Outer Space: On the Road to Debris Mitigation', *en United Nations Institute for Disarmament Research (ed.), Security in Space: The Next Generation*, 81–103. Geneva: United Nations.

Wolfe, Richard A. 1994. 'Organizational Innovation: Review, Critique and Suggested Research Directions.' *Journal of Management Studies* 31 (3): 405–431. https://doi.org/10.1111/j.1467-6486.1994.tb00624.x.

World Health Organization (WHO). 2000. *Towards an Assessment of the Socioeconomic Impact of Arsenic Poisoning in Bangladesh.* Geneva: WHO.

Wright, Gary A. 1969. *Obsidian Analyses and Prehistoric Near Eastern Trade: 7500 to 3500 BC.* Anthropological papers, Museum of Anthropology, University of Michigan. Ann Arbor: University of Michigan.

Xiaoye, Liang, Zhihua Wang, Zhujin Zhou, Zhenya Huang, Junhu Zhou and Kefa Cen. 2013. 'Up-to-Date Life Cycle Assessment and Comparison Study of Clean Coal Power Generation Technologies in China.' *Journal of Cleaner Production* 39: 24–31. https://doi.org/10.1016/j.jclepro.2012.08.003.

Xie, Shuangyu, and Kohji Hayase. 2007. 'Corporate Environmental Performance Evaluation: A Measurement Model and a New Concept.' *Business Strategy and the Environment* 16 (2): 148–168. https://doi.org/10.1002/bse.493.

Zaltman, Gerald, Robert Duncan and Jonny Holbek. 1973. *Innovation and Organizations.* New York: John Wiley & Sons.

INDEX

accountability
  human responsibility, 342
  space debris, 343
  SSN, 343
accountics
  abstraction process, 67
  accounting research, 67
    empiricism, 68
  accounting
    academic research, 69
  definitions, 66
  econometrics, 67
  empirical incvestigation, 69
  market conditions and economic realities, 70
  mathematical operators, 70
  stakeholders, roles, 71
accounting and accountability
  implicit and explicit overlaps, 35
    budgeting process, 37
    governance structure, 37
    practice in firms, 35
    standards, 36
  technical complexities, 183
accounting information system (AIS), 301
accounting science
  practice, 108
accounting theories, 102
  CSR, 137

  economic thinking, 138
  EMA, methodological improvements, 136
  firms
    environmental considerations, 136
  principles
    concern, 104
    conservatism, 103
    historical cost, 103
  SNA, 136
  stakeholders, 102
accounting
  environmental consideration, framing, 186
    accounting standards, 187
    demand, 186
    EMA and CSR, 187
  paradigms, 182
    SEA theories, 183
    sustainability, 183
  practitioners
    role, 107
accounts information lifecycle, 105
activity-based costing (ABC), 81
activity-based management (ABM), 81
allowable accounting units (AAUs), 285
American Society for Quality (ASQ), 309

Index

analytical hierarchical process (AHP), 352
anthropocene
  aboriginal societies, 11
  challenges, 6
  ecocentrism, 10
  ecological sciences, 9
  modern medicines, 6
  natural systems and resources,
    issues, 8
    signs, 8
  scientific studies, 9
  western scientific approach, 5

business failure
  co-citation analysis, 61
  dissolution cases, 57
  dissolution of firms, 59
  economic method and accounting focus, 57
  fund flow approach, 60
  inertia of similarities, 62
  liability
    newness, 62
    obsolescence, 62
    smallness, 62
  organisational learning theory, 62
  ratio based models, 61
  resource based theory, 62
  resource dependence theory, 62
  structural changes, 58
  technological adaptations, 63

cap-and-trade system, 91
carbon accounting, 162
  CDM, 163
  CER, 165
  ECEA, 166
  ETS, 163
  GCC, 162
carbon development project (CDP), 159
carbon disclosure project (CDP), 161

carbon management accounting (CMA), 168
carbon trade
  accounting implications, 164–167
case-based reasoning, 352
certified emission reductions (CER), 165
clean development mechanism (CDM), 163
coal combustion residuals (CCRs), 347
coal combustion waste (CCW), 346
cognitive limitations
  cost of waste, 221
  dynamic systems, 223
  ecological receptors, 223
  environmental impacts, layered nature, 224
complex waste
  mathematical modelling, 346
  CCW, 346
  fly ash (FA), 346
corporate accounting
  environmental care, role, 109
  cost accounting, 112–114
  financial accounting, 109–112
  LCC, 114–115
  managerial accounting, 115–117
  organisational externalities, 117
  reporting, 109–112
  traditional accounting practice, limitations, 117
corporate social reporting/responsibility (CSR), 137
corporate sustainability reports (CSR), 175
corporate sustainability responsibilities (CSR), 126
cost accounting, 73
  inventory accounting, generation
    FIFO, 75
    LIFO, 75
cost and managerial accounting, 72

cost benefit analysis (CBA), 81
cost-volume-profit (CVP) analysis, 82
costing, 73
   cost determination, 76
   cost planning, 76
   enterprise resource planning (ERPs), 79
   techniques, 76
      budgetary analysis, 76
      standard costing, 76
costs
   formulation, 74
   operations and transformation processes, 74

data envelopment analysis (DEA), 352
decision making
   information beyond numbers and interpretive elements, 83
   accounting methods, 84
   accounting shares, 84
   economic science, emphasis, 85
   issues handling, 85
   professional management, 84
   quantitative methods, emphasis, 85
decision-tree modeling
   traceability of higher order impacts, 360–355
   use, 351–353
dimensional view
   accounting universe
      environmental impacts, 189–191
      methodological improvisation, 192–194
      pragmatism, role, 189

eco-intensity change index (EICI), 292
ecological sciences, 333
   economics and accounting, 338–342

energy, 334
   mono-culturism, 337
   multi-dimensionality, 345
   natural degeneration, 344
   philosophical level, 337
   sustainability, 335
economic and environmental policy evaluation techniques
   sectoral and regional level, 349–351
economic science, 67
effluent treatment plant (ETP), 263
elimination and choice expressing reality (ELECTRE), 352
Emission Trading Scheme (ETS), 163
emissions, 162
   CDM, 163
   CER, 165
   ECEA, 166
   ETS, 163
   GCC, 162
enterprise resource planning (ERPs), 79
environment management system (EMS), 15, 307
   certification standards, 313
   complimentary nature and EA, 318–321
   differentiation needs, 310
   empirical validations, 312
   environmental care frameworks, multiplicity, 308
      ASQ, 309
      institutionalized, 309
      PDCA, 308
   EU-EMAS, 312
   organizational interaction, 311
environmental accounting (EA), 119, 188
   analysis of key findings, 245–247
   business transactions
      book-keeping environmental, schematics, 201–203
      carbon accounting, 271

Index

GHG, 273
LULUCF, 272
microeconomic, 272
supra-national level, 275
case study, 258
  accounting of environmental aspects, 258
  computation of environmental aspects, 258
  emissions and accounting, 263–264
  environmental care, 264–266
  sensitive decision-making, 266–267
  solid waste and disposal, 260–262
  valuation, 258
  waste water, 262–263
CHS, 247, 257
comparative positioning, 197
corporate accounting framework, systemic integration, 194–201
CSR, 175
eco-efficiency, 178
ecological/deep green view, 181
EMA methodologies, 226
  key findings, analysis, 229–232
  manufacturing environment, 227–229
  project site, 227
emission accounting, transportation, 281–285
emission flow, 276
  GHG, 276
  MFCA, 276
  unification, 277
emission trading, 271
  GHG, 273
  LULUCF, 272
  microeconomic, 272
  supra-national level, 275
emissions T-account, 254
environment and economics, 174–175

environmental aspects
  quantification, 237
  valuation, 237
environmental ledger of aspects, 239–228
environmental liability T-account, 255
experimental construct using FPP, validation, 279–281
externalities and accounting, cost, 225–213
FCA, 176
  techniques, 178
flow of resources, 232
green view, 179
instrumental view, 178
integrated waste, 276
  GHG, 276
  MFCA, 276
  unification, 277
LCA, 178
literature bridging, 177
methodological complexities, 176
process flow, 194
relevance, 204
service environment, 244–245
solid waste (externality) T-account, 252
sustainability, 205
theory of ethical rights, 179
timeliness, 204
traceability, 203
uniformity, 204
waste water T-account, 253
environmental capable enhancing asset (ECEA), 166
environmental cost accounting (ECA), 133
environmental impact assessment (EIA), 13
environmental management accounting (EMA), 124, 133, 139
  accounting methodology, 140–141
  FCA, 139

framework, 134
laws, 135
LCC, 139
literature, review, 134
methodological developments, 20
MFCA, 139
 waste accounting, 134
UNDSD, 134
 waste accounting methodology, 141–143
environmental management system (EMS), 199, 291
environmental performance indicators (EPIs), 134, 149–153
 firm level, 153–156
environmental performance industries (EPIs)
 eco-efficiency, 290
 EICI, 292
 EMS, 291
 environmental accounting, role, 294
 KLD performance rankings, 291
 WTP, 292
environmentally enhanced life cycle costing (E-LCC), 145
 CSR, 147
 FDF, 147

financial accounting, 50
 acid test, 50
 balance sheet view, 55
 Commission Statement, 51
 FASB and IFRS, 51
 firms, value realization, 53–55
 GAAP, 50
 IASC, 56
 income and profitability view, 55
 inter-company profit, 52
 issues, 53
Finnish Defence Forces (FDF), 147
firms
 accountability, 37
 business entities, 42–45

 concept of Dharma, 47
 debates, 42
 equation, 46
 ethical theories, 42–45
 institutionalization, 48
 organisational failures, 40
 principal agent relationship, 47
 roles of ethics, 40
 social learning process, 38
 societal expectations, 39
 transactional aspects, 38
fly ash (FA), 346, 347
 CCRs, 347
 components, 347
 ground water contamination, 348
 hidden costs, 359–360
 human, health issues, 349
 lifecycle and externalities, methodological complexities, 351
 soil contamination, 347
 water contamination, 348
framework(s), 295
 financial reporting, 296
 genesis of multiplicity, 297
 legitimacy, 297
 quantitative data, 299
 SEAR, 296
 sustainability, 301
full cost accounting (FCA), 134, 139, 147–149, 209
 energy consumption, 213
 GHG emissions, 213
 legal perspective, 217
 MSW, 210
  activities and environmental impacts, 210–212
 SWM services, 216
 techniques, 178
fuzzy set theory, 352

generally accepted accounting standards (GAAP), 50
global climate change (GCC), 162

Index

global reporting initiative (GRI), 167
  reporting framework, 167
  UNEP, 167
goal programming (GP), 352
green accounting, 119
  disclosure process, 129
    ethical rights theory, 132
    legitimacy theory, 130
    political economy theories, 129
    research in countries, 130
  firm level, 124–125
  regional/macro level, 120–124
  SEA, 125
    basic accounting function, 125
    CSR, 126
    ethical approach and rights view, 127
    instrumental view, 127, 128
    methodological view, 126
  SEAR, 127
greenhouse gases (GHG), 159
  accounting, 159–161, 169
greening firms, 314
  EMS framework, 314
greening information needs
  firms, 301–302
    information system, environment impacts, 303–306
greening supply chain management (GSCM), 314–316

high and low level radioactive wastes (HLRW and LLRW), 90

integrated reporting, 295
  financial reporting, 296
    genesis of multiplicity, 297
    legitimacy, 297
    quantitative data, 299
    SEAR, 296
    sustainability, 301
Integrated Waste Management (IWM) framework

bio-wastes, 325
CHS facilities, 323
ecotel certification, 326
employee trainings, 325
EMS certification, 321
environmental stewardship, 329
generalizations, 328
green produce, use, 325
organizational eco-friendly measures and positive impacts, 322
reduction in energy use, 324
solid wastes cycle, 323
structural elements, 323
waste water cycle, 324
International Accounting Standards (IAS), 165

joint implementations (JI), 163

land use and land use changes and forestry (LULUCF), 272
lean accounting
  advanced techniques, 82
life cycle costing (LCC), 114–115, 178

management accosting
  tools and techniques, 81
managerial accounting, 115–117
  ABC, 81
  ABM, 81
  CBA, 81
  core decision making areas, 80
  CVP, 82
  defined, 79
  performance measurement activities, 80
  SMA, 82
  theory and application, 80
manufacturing execution function (MES), 302
material flow accounting (MFA), 123

material flow cost accounting
(MFCA), 139
material flow cost analysis (MFCA),
143
   SMEs, 144
   studies, 144
monetary system, 4
multi criteria decision analysis
(MCDA), 220
   ecological modelling, 220
multi-attribute utility theory (MAUT),
352
multi-criteria analysis (MCA), 352
multivariate discriminant analysis
(MDA), 60
municipal solid waste (MSW), 210
   activities and environmental
impacts, 210–212
   SWM services, 216

Netherlands – national accounting
matrix including
environmental accounts
(NAMEA), 122

organisational activities
   environmental view, 87
      economic theories, 87
      societies, 87
      waste and emissions, 88
organizational accountability, 25
   business firms, evolution, 26
   firms, 26
      alternate and behavioural
school, 31–32
      economic theories, 26, 33–35
      justification, 26
      theory of profit/sales
maximization, 27–28
      transaction cost theory, 28–31

Perkins Framework, 122
physical input-output tables (PIOT),
122

Plan-Do-Check-Act (PDCA), 308
preference ranking organization
method for enrichment
evaluation (PROMTHEE),
352
proposed model
   theoretical evaluation, 355–356

quality costing, 78

sample business transactions
   comparative accounting
viewpoints, 195
simple multi-attribute rating
technique (SMART), 352
small and medium enterprise (SMEs),
144
social and environmental accounting
(SEA), 16, 20, 125, 125
   basic accounting function, 125
   CSR, 126
   ethical approach and rights view,
127
   instrumental view, 127, 128
   methodological view, 126
   SEAR, 127
social and environmental accounting
reporting (SEAR), 127, 296
Space Surveillance Network (SSN),
343
spatio-temporal insights, 119
strategic management accounting
(SMA), 82
sustainability, 7, 94, 182, 333
sustainability and accounting sciences
   accountability, 4
      sustainability, 5
      western form, 5
   contemporary positioning
      firm environment exchange, 16
      monetary implications, 15
      roles, 15
      economic gains, 11
      cases, 12–13

Index

ecocentric approach, 14
environment and accounting theories
  contemporary advances, 19–20
  environmental accounting dimensional view, 20–22
  ontological origin, 18
  theories
    convergence/divergence, 22–23
  trade and commerce, 1
  traditional paradigm, 18–19
sustainability reporting, 167
  framework, 167
  UNEP, 167
sustainability
  aboriginal societies, 11
  challenges, 6
  ecocentrism, 10
  ecological sciences, 9
  environmental aspects, 97
  micro-level application, 177
  modern medicines, 6
  natural systems and resources,
    issues, 8
    signs, 8
  scientific studies, 9
  SEA theories, 183
  waste
    decomposition, 95
    emission, increasing level, 95
  western scientific approach, 5

system of ecological and environmental accounting (SEEA), 121
system of national accounts (SNA), 119, 121, 136

target costing, 77
theory of constraints (TOC), 77
total cost approach (TCA), 134, 218
  fly ash (FA), generalize externalities, 353–354
  MCDA, 220
  modelling, 218
traditional accounting techniques, 117
triple bottom line (TBL), 159

United Nations Environment Programme (UNEP), 167
units of measurement (UOMs), 199

waste ownership, 89
  cap-and-trade system, 91
  disposable mechanism, 89
  externalities, 91
    accounting practices, 93
    carbon dioxide (CO_2), 92
    green house gases (GHGs), 92
    transferring, 93
  HLRW and LLRW, 90
  legal jurisprudence, 90
willingness-to-pay (WTP), 292

ABOUT THE AUTHOR

Somnath Debnath received his doctorate from the Birla Institute of Technology, India, for his contribution to incorporating environmental considerations in managerial accounting. He also holds a master's degree in business administration (MBA) from Walden University, USA, and a master's in technology (MTech) from Swinburne University of Technology, Australia. Debnath is a Fellow of the Institute of Cost and Management Accountants of India, being a qualified Cost and Management Accountant (CMA), and an academic member of the Athens Institute for Education and Research, Greece, and also of the International Engineering and Technology Institute, Hong Kong. He is a senior solutions architect at Zensar Technologies and a former managing principal at Oracle Corporation.

Debnath's academic interests have been to advance environmental thinking in the field of accounting and in the information sciences, and to support the managements' need for green information beyond the prevailing business constructs, in turn to enable firms to handle sustainability challenges. This includes greening different business functions. His research interests include diverse areas of business such as the decision sciences, project management, systems analysis and design, requirements engineering, green information systems and grey mathematics. He is also a reviewer of a number of academic journals.

In addition, Debnath is a technology expert in the ERP technology space—a subject expert, architect, functional consultant and project manager rolled into one—who has consulted for top Fortune® 500 organizations in the last two decades and has supported the industry with business process reengineering and automation needs. To this end, he has been a part of more than 10 full life-cycle ERP implementations and rollouts in different capacities. His latest area of expertise is in cloud-enabled ERP and setting up cloud application centres of excellence for information systems vendors.